SpringerWienNewYork

CISM COURSES AND LECTURES

The series presents lecture notes, monographs, edited works and proceedings in the field of Mechanics, Engineering, Computer Science and Applied Mathematics.
Purpose of the series is to make known in the international scientific and technical community results obtained in some of the activities organized by CISM, the International Centre for Mechanical Sciences.

INTERNATIONAL CENTRE FOR MECHANICAL SCIENCES

COURSES AND LECTURES - No. 500

DYNAMICAL SYSTEMS, WAVE-BASED COMPUTATION AND NEURO-INSPIRED ROBOTS

EDITED BY

PAOLO ARENA
UNIVERSITY OF CATANIA, ITALY

SpringerWienNewYork

This volume contains 126 illustrations

Printed in Italy
SPIN 12244348

All contributions have been typeset by the authors.

ISBN 978-3-211-78774-8 SpringerWienNewYork

PREFACE

This volume is a special Issue on "Dynamical Systems, Wave-based computation and neuro-inspired robots", *based on a Course carried out at the CISM in Udine (Italy), the last week of September, 2003. From the topics treated within that Course, several new ideas were formulated, which led to a new kind of approach to locomotion and perception, grounded both on biologically inspired issues and on nonlinear dynamics. The Course was characterised by a high degree of multi-disciplinarity. In fact, in order to conceive, design and build neuro-inspired machines, it is necessary to deeply scan into different disciplines, including neuroscience, Artificial Intelligence, Biorobotics, Dynamical Systems theory and Electronics. New types of moving machines should be more closely related to the biological rules, not discarding the real implementation issues. The recipe has to include neurobiological paradigms as well as behavioral aspects from the one hand, new circuit paradigms, able of real time control of multi joint robots on the other hand. These new circuit paradigms are based on the theory of complex nonlinear dynamical systems, where aggregates of simple non linear units into ensembles of lattices, have the property that the solution set is much richer than that one shown by the single units. As a consequence, new solutions "emerge", which are often characterized by order and harmony. Locomotion in livings is a clear example of this concept: ordered motion is the solution of a great amount of concurrently co-operating neurons; neural "computation" is also rather "wave based", than "bit based". In this direction, continuous time spatial temporal dynamical circuits and systems are the paradigmatic mirror of neural computation.*
The volume mainly reflects the structure of the Course, but is directed toward showing how the arguments treated in that CISM Course were seminal for the subsequent research activity on action-oriented perception. The volume is therefore constituted of three main parts: the first two parts are mainly theoretical, while the third one is practical. The theoretical aspects, reported in the first part of the volume, discuss new programmable processing paradigms, the Cellular Nonlinear Networks (CNNs). These architectures constitute wave based computers processing spatial-temporal flows. Another important theoretical topic regards neurobiological and neurophysiological basis of in-

formation processing in moving animals, with the introduction of the paradigm of the Central Pattern Generator. Then a unifying view will be presented, where CNN approach, the neurobiological aspects and the robotic issues will be organically fused together, referring to a number of bio-inspired robotic prototypes already developed and really working. Finally, very interesting issues regarding how bio-robots can be used to model biological behavior are given, together with examples of neural controllers based on spiking neurons, applied to model optomotor reflex and phonotaxis.
The second part of the volume includes the use of sensory feedback in locomotion controlled by the CPG. Then a looming detector for collision avoidance, inspired by the locust visual system, is modelled and shown. Sound localization and recognition is also addressed by using a network of resonate and fire neurons and finally a chapter introduces the main aspects of robot perception.
School attendees were also allowed to implement and realise some applications of what theoretically learned, helped by a series of practical tutorials introduced by Tutors. The results of the practical work done by the students have been also added at the end of the volume to demonstrate the interest shown by attendees and the implementation of the new ideas conceived by them.
A special thank goes to Prof. M. G. Velarde, for supporting the organization of the Course. The Coordinator particularly thanks Prof. Leon O. Chua for transmitting the CNNs basics, starting point of a large research wave. The invited speakers Prof. B. Webb and Prof. T. Deliagina are acknowledged for their contribution through interesting and attracting lessons and for collaborating in the production of the present volume. A warm thank is also addressed to Dr. M. Frasca and Dr. A. Basile, who actively worked as Tutors for the students during the practical part of the Course. The Coordinator would like to thank Prof. L. Fortuna for transmitting the view of nonlinear complex dynamics within multidisciplinary research, giving rise to a large part of the contents of the volume and to the following research activity on robot perception.

Prof. Paolo Arena
Dipartimento di Ingegneria Elettrica Elettronica e dei Sistemi,
Universitá degli Studi di Catania, Catania, Italy

CONTENTS

Part I

Foundations of Neurodynamics and wave based computation for locomotion modeling

Overview of Motor Systems. Types of Movements: Reflexes, Rhythmical and Voluntary Movements

Tatiana G. Deliagina

Department of Neuroscience, Karolinska Institute,
SE-171 77, Stockholm, Sweden

Abstract One of the principal characteristics of the animal kingdom is the ability to move actively in space. Our movements are controlled by a set of motor systems that allow us to maintain posture, to move our body, head, limbs and eyes, to communicate through speech. Motor control is one of the most complex functions of the nervous system. During movement, dozens and even hundreds of muscles are contracting in a coordinated fashion. This coordination is a basis for a remarkable degree of motor skill demonstrated by dancers, tennis players and even by ordinary people when walking or writing a letter.

1 Types of movements

Most movements performed by animals and humans can be divided into three broad classes: reflex responses, rhythmical movements and voluntary movements.

Reflexes are relatively rapid, stereotype, involuntary responses that are usually controlled in a graded way by a specific eliciting stimulus. For example, protective skin reflexes lead to withdrawal of the stimulated part of the body from a stimulus that may cause pain or tissue damage. Coughing and sneezing reflexes remove an irritant from the nasal or tracheal mucosa by inducing a brief and strong pulse of air. This is caused by synchronized activation of abdominal and respiratory muscles triggered by afferents activated by irritant. Swallowing reflexes are activated when food is brought in contact with mucosal receptors near the pharynx. This leads to a coordinated motor act with sequential activation of different muscles that propel the food bolus through the pharynx down the esophagus to the stomach. Postural reflexes are responsible for maintenance of the body and its parts in a stationary position.

Rhythmical movements are characterized by sequence of relatively stereotyped, repetitive cycles generated automatically. For example, we are continuously breathing from the instant of birth, without thinking about each inspiration . expiration movement. In contrast to breathing, the majority of rhythmical movements are not generated continuously but should be initiated either voluntary (like locomotion in higher vertebrates) or by specific sensory stimuli (like scratching in cats or dogs or locomotion in invertebrates).

Voluntary movements. Examples of this wide class of movements are the skilled movements of fingers and hands, like manipulating an object, playing the piano, reaching, as well as the movements that we perform in speech. Voluntary movements are characterized by several features. They are purposeful, goal directed, initiated in response to specific external stimuli or by will. The performance of voluntary movements improves with practice. As these movements are mastered with practice, they require less or no conscious participation. Thus, once you have learned to drive a car you do not think through the actions of shifting gears or stepping on the brake before performing them. It is necessary to note, however, that this classification of movements is not perfect because it is difficult to draw a clear-cut dividing line between the different classes. Voluntary movements with practice become more and more automatic like reflexes. By contrast, rhythmical movements and reflexes can also be modified by will. For example, we can voluntary terminate our rhythmical breathing movements when diving, we can also modify the duration of the inspiration and expiration when singing. If necessary, we can keep a hot object in our hand despite it can damage the skin. This is possible because of voluntary inhibition of protective withdrawal reflexes. If, however, the object is touched without knowing that it is hot, the hand will be withdrawn automatically with the shortest possible latency. Despite we define reflexes as stereotyped movements, some reflexes can underlay plastic changes. A good example is the vestibulo-ocular reflex. This reflex is responsible for stabilization of the visual image on retina during head movements. For instance, movement of the head to the left evokes movement of the eyes to the right with such a speed and amplitude that the visual image on retina does not move. The movement of eyes is initiated by the signal from vestibular afferents activated by head movement. When the animal observes the world through minifying or magnifying glasses (that alter the size of visual image on the retina) the compensatory eye movements, that would normally have maintained a stable image of an object, are now either too large or too small. Over time, however, the vestibulo-ocular reflex recalibrates and the amplitude of eyes movement changes in accordance with the artificially altered size of the visual field. Finally, different rhythmical movements, from the point of view of their initiation, do not represent a homogeneous group, since some of them are initiated as reflexes (like scratching, paw shaking), but others, like locomotion in higher vertebrates, are initiated voluntary.

2 Basic components of motor system

2.1 Motoneuron

A movement is performed due to a contraction of muscles which, in their turn, are controlled by motoneurons. Each motoneuron sends its axon to one muscle and innervates limited number of muscle fibers. A motoneuron with its muscle fibers is referred to as a motor unit, since a single action potential generated by the motoneuron evokes contraction of all muscle fibers that it innervates. All motor commands eventually converge on motoneurons, whose axons exit the CNS to innervate skeletal muscles. Thus in Sherriningtons words, the motoneurons form a "final common pathway" for all motor actions.

2.2 Neuronal networks generate motor patterns; Central pattern generators

Each type of motor behavior can be characterized by its own motor pattern, which can be defined as the sequence and degree of activation of particular muscles. For example, the locomotor pattern consists of alternating activity of flexor and extensor muscles around different joints of the limb, with specific phase shifts between different limbs. Each of the numerous motor patterns is generated by a group of neurons. the neuronal network. The network contains the necessary elements and information to coordinate a specific motor pattern such as swallowing, walking, breathing. When a given neuronal network is activated, the particular motor pattern is expressed. A typical network consists of a group of interneurons that activate a specific group of motoneurons in a certain sequence. The interneurons also inhibit other motoneurons that may counteract the intended movement. For normal functioning of the network, sensory information signaling about execution of a movement is usually very important. In many cases, however, the network can generate the basic motor pattern without sensory feedback, though the pattern more or less differs from the normal one and is not adapted to the enviroment. Such networks are often referred to as central pattern generators (CPGs). There are CPGs for locomotion, scratching, swallowing, breathing, etc., which can be activated in in vitro, immobilized, or deafferented preparations in which sensory feedback is absent.

2.3 Command systems

Each particular motor network operates when it is activated. This function is performed by command systems. A command system integrates different sensory and central signals, and sends a command to the corresponding network. In response to this simple command, the network generates a complex motor pattern. A good example is the command systems for initiation of locomotion. Tonic activation of neurons of the mesencephalic locomotor region (MLR) in the brainstem evokes coordinated locomotor movements of limbs due to activation of spinal locomotor network.

2.4 Role of sensory information in movement control

Sensory contribution to motor control is very important in most types of movements. Sensory information from different receptors is used by the motor systems in a number of ways:

1. Specific sensory signals can trigger behaviorally meaningful motor acts, that is initiate them as reflexes. For example, the signals from mucosal receptors located near the pharynx and activated by contact with food, evoke swallowing reflex. The signals from skin receptors located on the head and trunk and activated by parasite insects, evoke scratching reflex.
2. Sensory signals can contribute to the control of an ongoing rhythmical movement by influencing the switch from one phase of the movement to another, as well as by affecting some other characteristics of the motor pattern. For instance, in breathing movements, sensory signals from the lung volume mechanoreceptors determine when inspiration is terminated. In locomotor movements, sensory information from muscle receptors about the hip position and the load on the limb affects the duration of stance and swing phases of the step cycle. In scratching movements, when all sensory inputs from the moving limb of the cat are eliminated by deafferentation (transection of dorsal roots of the spinal cord containing sensory fibres), the limb still is able to perform rhythmical movements, but these movements cannot reach the goal . to remove the irritant from the skin, because they are performed without toughing the skin.
3. A great variety of sensory signals, which provide information about the position of different parts of the body in relation to each other and to the external world, are very important for the generation of voluntary movements. For example, to perform a reaching movement

(to move a hand toward a specific object) it is necessary to know the initial position of the hand. If the hand is located to the left or to the right of the object, different motor patterns should be generated to bring the hand to the target.

4. Sensory signals are used for corrections of the perturbed movement or posture. For instance, hitting an obstacle during walking causes activation of skin afferents. These sensory signals evoke a limb extra flexion, which is incorporated into the swing phase of the locomotor cycle. As a result, the movement is corrected without termination of walking. Acceleration of the bus leads to a disturbance of the vertical body orientation in passengers (body sway in the opposite direction). This causes activation of a number of sensory systems. These sensory signals evoke postural corrective response, that is coordinated contraction of specific muscles of the legs and trunk, which return the body to the vertical position. These examples represent feedback principle of motor control, that is a compensation for the actual perturbation when it has occurred. A limiting factor for the efficacy of feedback control in biological systems is the delay involved. A sensory afferent signal must first be elicited in the receptors concerned. It then has to be conducted to the nervous system and be processed there to determine the proper response. The correction signal must subsequently be send back to the appropriate muscle(s) and make the muscle fibers build up the contractile force required. For example, in humans it may take several hundred milliseconds to respond to visual cues, while a quick (for example reaching) movement itself may last only for 150-200 ms. That is why the feedback mechanisms can be used only to control slower movements or to maintain a posture, that is to stabilize a certain body orientation in space.

5. Sensory signals can be used for anticipation of expected disturbance of movement or posture. For example, if during locomotion the cat sees the obstacle, this visual information is used for modification of the locomotor cycle (generation of extra flexion during swing phase) at the moment when animal reaches the obstacle. This results in overstepping the obstacle. Thus, a perturbation is anticipated before it is initiated, and correction begins before the perturbation has actually occurred. This principle of motor control is called feed forward control. Usually in motor systems the feedback control supplements the feed forward control.

2.5 Development of motor systems

The neuronal networks that allow performance of the basic movement repertoire (e.g. locomotion, posture, breathing, eye movements, etc.) as well as the networks that underlie reaching hand and finger movements, sound production as in speech, are genetically predetermined and constitute the motor infrastructure that is available to a given individual after maturation of the nervous system has occurred. Motor systems develop through maturation of the neuronal substrate and by learning through different motor activities. In development of reflexes and rhythmical patterns, the process of maturation plays the primary role. By contrast, in development of voluntary movements, learning through playing represents an important element both in children and in young mammals such as kittens and pups. Different animals are born at different degree of maturation of their motor systems. Human infant is comparatively immature and has very limited behavior repertoire. It is able to breath, and has searching and sucking reflexes so that it can be fed from the mother.s breast. It can swallow and process food. A baby also has a variety of protective reflexes that mediate coughing, sneezing, and limb withdrawal. These different patterns of motor behavior are thus available at birth because their networks are mature already at birth. During the first year of life the human infant matures progressively. It can balance its head at 2-3 month, is able to sit at around 6-7 months, and stand with support at approximately 9-12 months. This development represents to a large degree a maturation process following a given sequence. In common language the child is said to .lean. to sit, to stand, to walk but in reality a progressive maturation of the nervous system is taking place. Identical twins start to walk essentially at the same time, even if one has been subjected to training and the other has not. In the beginning, the locomotor pattern is very immature. Proper walking coordination followed by running appears later, and the basic motor patterncontinues to develop until puberty. The fine details of the locomotor pattern are adapted to surrounding world, but also can be modified by will. While the newborn human is comparatively immature, some other mammals, such as horses and deer, represent another extreme. The young calf of the antelope gnu can stand and run directly after birth. So, the neuronal networks underlying locomotion, equilibrium control and steering must be sufficiently mature and available at birth. In addition to the reflexes and rhythmical movements, humans also develop voluntary movements - skilled motor coordinations, allowing delicate hand and finger movements to be used in handwriting or playing an instrument or utilizing the air flow and shape of the oral cavity to produce sound as in speech or singing. The neural substrates allowing learning and execution of these complex motor sequences are expressed genetically. But particular motor coordinations are learned, however, such as which language one speaks or the type of letters one writes. For instance, when learning to play the flute, the particular finger settings that produce a given tone and repeats many times can be retained in memory, along with the sequence of tones that produce a certain melody. Thus particular muscle combinations that produce a sequence of well-timed motor patterns are stored. It is characteristic of a given learned motor pattern that one can "call" upon it to perform a given motor act over and over again in rather automatic way.

2.6 Distribution of motor functions in CNS

hierarchically and in parallel. The neuronal networks responsible for the generation of basic motor coordinations (reflexes, rhythmical movements) are located at lower levels of CNS - in pedal ganglia (mollusks), in segmental ganglia (insects), and in the spinal cord and brainstem (vertebrates) (see Figure 1). They are activated by the commands arriving from higher levels of CNS. In mollusks, command neurons are located in cerebral ganglia; in insects they reside in cerebrum. In vertebrates, these commands originate from motor centers of the brainstem and motor areas of cortex, they are transmitted by different descending pathways (see Figure 2). The motor centers of the brainstem are in turn under control of the higher level centers responsible for selection of motor behaviors.

The same movement can performed in context of different motor behaviors. For example, the animal can locomote during migration, escape reaction, hunting, etc. In these cases, the same motor network is activated through different command systems.
Sensory signals used for feedback control of movements are processed in different parts of CNS in parallel. For example, sensory signals from the stepping limb during locomotion enter the spinal cord and affect the interneurons and motoneurons of the spinal locomotor network directly, and also indirectly, through spino-cerebellar loop (see Figure 2).
The spinal cord is the lowest level of motor control in vertebrates. Its motor capacities are studied in the animals with a transection of the spinal cord. If the cord is cut in the upper (cervical) region, the animal is not able to breath, and ventilation of its lungs should be performed artificially. It also cannot maintain the upright body posture. But if the body is supported, and specific stimuli are applied, spinal animals can perform a number of movements (spinal reflexes) such as the flexion and crossed extension reflexes. The spinal cord also contains the neuronal networks for generation

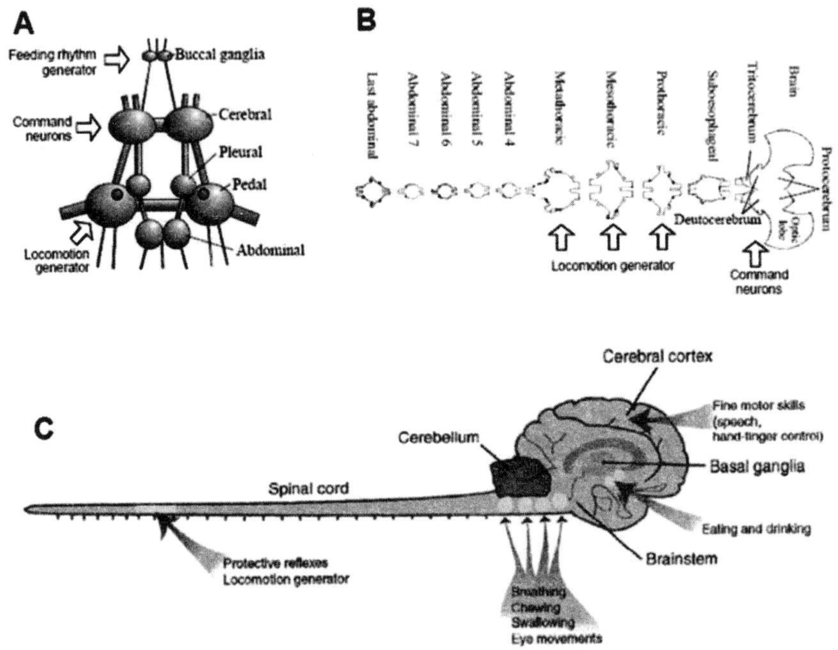

Figure 1. Localization of some motor functions in CNS of different species. (A) CNS of mollusk Clione contains 5 pairs of ganglia. The locomotion generator is located in pedal ganglia, the feeding rhythm generator. in buccal ganglia. They are activated by command neurons from cerebral ganglia. (B) CNS of locust. The locomotion generator is located in thoracic ganglia, and activated by command neurons from the brain ganglia. (C) CNS of a higher vertebrate animal. Most motor networks are located in the spinal cord, brain stem, and motor cortex.

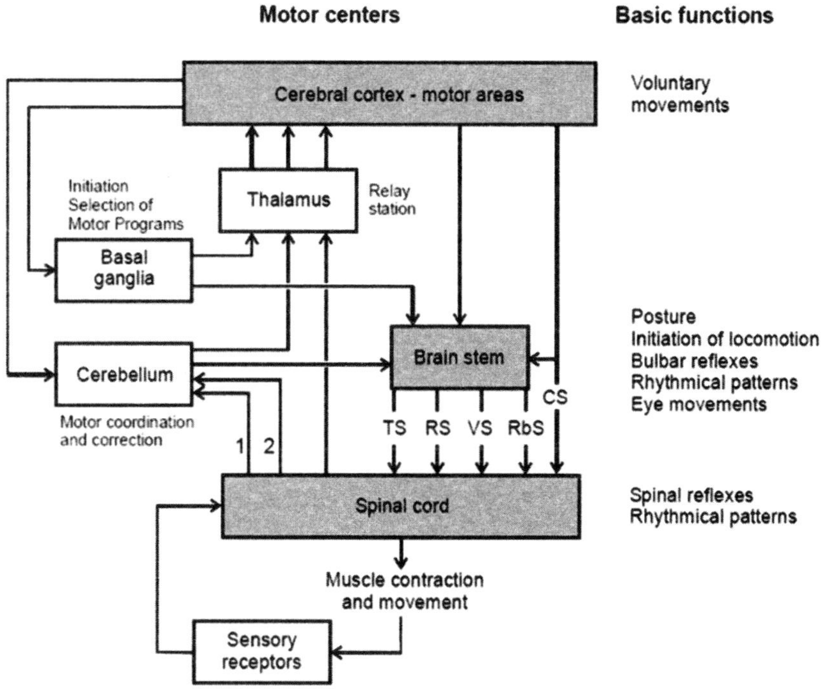

Figure 2. Main motor centers of CNS, their relationships, and basic functions. Abbreviations: CS, TS, RS, VS, RbS – cortico-, tecto-, reticulo-, vestibulo-, and rubro-spinal descending pathways. 1, 2 – sensory and central feedback signals coming to cerebellum.

of such rhythmical movements as scratching and locomotion. In the spinal dog,stimulation of the skin on its back and sides evokes scratching movements. In the spinal cat positioned on the moving belt of the treadmill, unspecific sensory stimulation (like mechanical stimulation of the tail base) evokes stepping movements. In intact animals, locomotion is initiated voluntary, and the spinal locomotor network is activated by the commands arriving from the higher levels of the CNS. By contrast, in spinal animals, the locomotor network can be activated by unspecific sensory stimuli.

The brainstem represents the second level of motor control. It contains a number of networks for generation of bulbar reflexes (swallowing, vestibulo-ocular reflex, coughing), eye movements, and rhythmical movements (breathing, chewing). In addition, the brainstem through descending pathways controls the movements generated by the spinal cord. An important function of the brainstem is integration of sensory information of different modalities - somatosensory, vestibular and visual. The brainstem also contains a number of specific motor centers like MLR, and the centers for regulation of the muscle tone.

The capacity of the brainstem-spinal mechanisms for motor control are clearly seen in the decerebrate animals (that is the animals with the brain transection between the brainstem and forebrain). These animals are able to maintain the basic (upright) body posture, to perform different types of locomotion (walking, trotting, galloping), to breath and adapt respiration to the intensity of movements, to swallow when food is put in the mouth. However, decerebrate animals perform these movements in a stereotyped fashion like a robot. The movements are thus not goal directed and poorly adapted to the environment, but otherwise coordinated in an appropriate way.

In contrast to decerebrate animals, decorticated animals (in which the cerebral cortex was removed), demonstrate a surprisingly large part of the normal motor repertoire, including some aspects of goal-directed behavior. They move around, eat and drink spontaneously. They can also learn where to obtain food, and search for food when hungry. They may also display emotions such as rage, and attack other animals.

The diencephalon and subcortical areas of telencephalon of the forebrain contain two major structures important for motor control: the hypothalamus and the basal ganglia. Hypothalamus is composed of a number of nuclei that control different autonomic functions, including intake of fluid and food (see Figure 2). The latter nuclei become activated when the osmolarity is increased (fluid is needed) or the glucose level becomes low (food is required). Continuous activation of paraventricular nucleus by electrical stimulation or local ejection of a hyperosmolaric physiological solution evokes a recruitment of a sequence of motor acts that appear in a logical order. The animals first starts looking for the water, then starts walking toward the water, positions itself at the water basin, bends forwards, and starts drinking. The animal will continue drinking as long as the nucleus is stimulated.

Basal ganglia are of critical importance for the selection and normal initiation of motor behavior. The output neurons of basal ganglia are inhibitory and have a very high level of activity at rest. They inhibit a number of motor centers in diencephalons and brainstem and also influence the motor areas of cerebral cortex via thalamus. When a motor pattern, such as a saccadic eye movement, is going to be initiated, the basal ganglia output neurons (that are involved in eye motor control) become inhibited. This means that the tonic inhibition produced by these neurons at rest is removed, and saccadic motor network in mesencephalon is relieved from tonic inhibition and becomes free to operate and induce an eye saccade to a new visual target.

It is very difficult for a casual observer to see the difference between a normal animal moving around in a natural habitat and a decorticated animal. It is only when specific tests are performed that one can see that the animal is lacking the skilled manipulations of the environment, such as picking fine food objects from small holes or fine foot placing during walking along the ladder.

Judging from experiments on primates, and patients that have suffered focal lesions of the frontal lobe, the cortical control of movements is of a particular importance for dexterous and flexible motor coordination, such as the fine manipulatory skills of fingers and hands and also speech. In the frontal lobe there are several regions (motor areas) that are involved directly in execution of different complex motor tasks, such as skilled movements used to control hands and fingers when writing, drawing, or playing an instrument. These different regions are organized in a somatotopic fashion. In the largest area, referred to as the primary motor cortex, areas taken up by the hands and the oral cavity are very large in humans, and are much larger than that for the trunk. This is explained by the fact that speech and hand motor control require a great precision and thus a larger cortical processing area than the trunk. The later is important for postural control but is less involved in the type of skilled movement controlled by motor cortex.

The cortical control of movement is executed in part by the direct corticospinal neurons (forming corticospinal tract, see Figure 2) but also by cortical fibers that project to brainstem nuclei from which descending pathways of the brainstem originate (such as rubrospinal, reticulospinal, vestibulospinal, etc). In addition, there are direct projections from cerebral cortex to the input area of the basal ganglia. The cortical control of motor coordination is thus achieved through both direct action on the spinal and brain stem motor centers but also to a significant degree by parallel action on a variety of brainstem nuclei.

Integrity of cerebellum is not necessary for the ability to generate movements, but lesions of cerebellum lead to reduction of their quality drastically. So, cerebellum is involved in coordination of movements. It receives inputs which not only carry sensory information about ongoing movements in all different parts of the body, but also information from different motor centers about intended (planed movements) even before a movement has been executed. The cerebellum also interacts with practically all parts of the cerebral cortex. This means that it is updated continuously about what goes on in all parts of the body with regard to movement and also about the movements that are planned in the immediate future. It was shown that integrity of the cerebellum requires for some cases of motor learning, for example, for recalibration of vestibulo-ocular reflex caused by environmental change.

In the following lectures, we will consider in more detail how the basic principles of motor control are realized in different motor systems. Special attention will be given the systems controlling locomotion and maintaining body posture.

Initiation and generation of Movements: 1. Central Pattern Generators

Tatiana G. Deliagina

Department of Neuroscience, Karolinska Institute, SE-171 77, Stockholm, Sweden

Abstract The complexity of motor control is, to a great extent, overcome by hierarchical organization of the controlling system. Lower levels of this system contain a set of central pattern generators (CPGs). the neuronal networks capable of producing the basic spatio-temporal pattern underlying different "automatic" movements (rhythmic movements like locomotion, respiration, as well as non-rhythmic ones like swallowing and defense reactions) in the absence of peripheral sensory feedback. Instead of controlling individual muscles involved in generation of a definite motor pattern, higher centers (through command system) activate the corresponding CPG that generates this pattern. The most detailed analysis of CPGs has been performed for rhythmical movements. In these experiments, sensory feedback was abolished using in vitro (see Figures 1D; 2D; 3D), immobilized (see Figure 4B-D), or deafferented preparations.

To figure out how a CPG operates one has to address the following questions: First, what is the source of rhythmicity in the network? Second, what mechanisms determine the temporal pattern of the motor output, that is, its frequency and the relative duration of the cycle phases? Third, what mechanisms shape the motor output, that is determine the number of phases in the cycle and the transition from one phase to another?

In the majority of CPGs, two parts can usually be distinguished: a rhythm generator and an output stage. The rhythm generator is the neuronal network in which the rhythm originates; it also determines a relative duration of the cycle phases. This network is usually formed by interneurons and does not include motoneurons. The output stage is formed by interneurons and motoneurons; they receive inputs from the rhythm generator but do not affect the rhythm. The output stage produces a final shaping of the motor output.

1 Origin of rhythmical activity

Recent evidence suggests that the rhythmic activity of many CPGs is based primarily on the endogenous rhythmicity (pacemaker properties) of generator neurons. Rhythmical activity persists in the generator neurons after surgical or pharmacological elimination of interactions between them. For instance, the generator neurons in the feeding CPGs of mollusks Planorbis and Clione, as well as the neurons of the swimming CPG of Clione continue to fire rhythmically after they have been extracted from the CNS (see Figures 1I-J; 2H-M). In the newborn rat, the generator neurons of the respiratory CPG in medulla continue to burst rhythmically after all synaptic interactions are blocked by a $low - Ca^{2+}/high - Mg^{2+}$ solution. In the lamprey spinal cord treated by tetrodotoxin (TTX) to block synaptic interactions, the generator neurons exhibit rhythmic membrane potential oscillations with a frequency typical of swimming (see Figure 3C).

The generator neurons from the respiratory CPG of vertebrates, as well as those from the Clione locomotor CPG, can be regarded as constitutive oscillators. At a certain level of the membrane potential, they exhibit rhythmic activity in the absence of external synaptic inputs (eliminated pharmacologically or by extraction of the neurons from CNS) and in the absence of any conditional factors. By varying the level of membrane potential in constitutive oscillators, the higher levels of CNS (through command systems) can easily control their oscillatory properties, and thus switch on and off the rhythmic oscillations and regulate their frequency (see Figures $1J_3$; 2I-K,M,N).

In contrast to constitutive oscillators, generator neurons in some CPGs express endogenous rhythmic activity only under the influence of some conditional factors. They are called *conditional oscillators*. For example, the STG generator neurons extracted from the ganglion together with their processes do not generate rhythmic activity at any membrane potential in the absence of the conditional factor; their rhythmic activity is triggered by pilocarpine. In the spinal cord, the TTX-resistant membrane potential oscillations in presumed neurons of the locomotor generator can be triggered by excitatory amino acids (see Figure 3C), which activate the locomotor CPG in all vertebrates. Thus conditional factors transform conditional oscillators from a passive to an active (rhythm-generating) state.

In addition to command inputs that activate CPGs, there also exist inputs that produce prompt termination of the CPG activity. These inputs can be defined as inhibitory command inputs. Some inhibitory inputs simply hyperpolarize generator neurons. In other cases, inhibitory inputs modulate the intrinsic membrane properties of generator interneurons. Their effect

results from the suppression of the ability of pacemaker neurons to produce an endogenous rhythm.
Although endogenous oscillatory properties of pacemaker neurons are the main source of rhythmogenesis in most CPGs, interactions between the neurons assist in reinforcing rhythmicity in the systems of mutually inhibitory groups of neurons. For example, the pacemaker property of generator neurons in locomotor CPG in Clione is not the only basis for the rhythm generation: the generator can produce the rhythm even when the interneurons are below the threshold of the pacemaker activity of individual neurons. In this case, a postinhibitory rebound is of critical importance: both group 7 and group 8 cells are capable of generating a single action potential after they have been released from inhibition (see Figure 1K). This is why, by producing an IPSP in the antagonistic group of neurons, a given group will evoke activity of these neurons on the "rebound" after termination of the IPSP (see Figure 1D).
In most CPGs, the endogenous rhythmic activity of pacemaker neurons and synaptic interactions reinforce one another, ensuring reliable rhythm generation. However, in some cases (in the locomotor CPGs of leech and mollusk Tritonia) endogenous pacemaker properties of the generator neurons were not found. In these CPGs, the rhythm generation is supposed to be based on the interactions between the generator neurons, that is on the network properties.

2 Formation of temporal pattern

A temporal pattern of the CPG output - the frequency of oscillations and relative duration of the cycle phases is, to a great extent, determined by the intrinsic properties of the generator neurons. The frequency of rhythmic activity produced by physically or functionally isolated generator pacemaker neurons is within the rage typical of a given movement, whereas the duration of membrane potential oscillations is comparable with the phase duration. In Clione, the pacemaker generator neurons produce prolonged ($\sim 150\ ms$) action potentials and, correspondingly, prolonged effects onto target neurons and thus contribute to determining the swimming phase duration (see Figure 1F-H). In young Clione, which generate higher frequency wing oscillations, both the interneuron action potentials and the phase duration are much shorter. In the feeding CPG of snails, the duration of retractor phase of the cycle is determined by the duration of endogenously generated plateau potentials in the generator interneurons (see Figure 2M,N).
In addition to cellular (intrinsic) properties of CPG neurons, their synaptic interactions also contribute to formation of the temporal pattern of the

generated rhythm. For a number of species (Clione, frog embrio, lamprey), it was shown that their locomotor CPG consists of two half-centers with mutual inhibitory connections. When these connections are abolished, the cycle period (generated by each isolated half-center) is shorter than in the presence of inhibitory connections (see Figure 3I,J). In the Planorbis feeding CPG, isolated protractor interneurons have intrinsic mechanism for burst termination (see Figure 2I-K). In the intact CPG, however, their discharges are terminated by an inhibitory input from the retractor interneurons (see Figure 2Q).

In some species (lamprey, frog embrio), which swim due to lateral undulatory movements of their body, the mechanism for generation of the whole locomotor pattern is composed of a chain of coupled segmental CPGs. The rhythm in this chain is determined by the fastest CPG (i.e. one with the highest frequency), whereas the direction of wave propagation is determined by its location (see Figure 3D-G). However, in the leech (which swims due to dorso-ventral undulatory movements of the body) the direction of wave propagation is determined by polarized connections between the segmental generators.

3 Shaping of motor output

The shape of the motor output (that is the number of phases in the cycle, the transition from one phase to the other, etc.) is determined primarily by the pattern of synaptic interactions between the CPG neurons. For example, the biphasic pattern produced by swimming CPGs in different animals is largely determined by excitatory connections between synergistic neurons (that is ones firing in the same phase), and by inhibitory connections between antagonistic ones (see Figures 1G; 3H). The three-phase output of the snail feeding CPG is determined by a more complex organization of intercellular connections. In this CPG (in contrast to swimming CPGs), connections between antagonistic neurons are asymmetrical. Protractor neurons excite retractor ones, which in turn inhibit protractor neurons (see Figure 2Q).

In addition to the connections determining the basic pattern of the motor output, there exist assisting mechanism that contribute to the reliable transition from one phase of the cycle to the other one. One of these mechanisms is a postinhibitory rebound, whereby the excitation of generator neurons of a given phase is facilitated after their release from inhibition in the previous phase. Another mechanism is delayed excitatory influences between the antagonistic groups of generator neurons, existing in parallel with their mutual inhibition. In lampreys and rats, the antagonistic half-centres of

the locomotor CPG fire not in succession but synchronously when crossed inhibition was blocked by strychnine. This suggests the existence of weak excitatory interconnections, masked by inhibitory connections, between the antagonistic half-centers. Excitatory interconnections might facilitate the transition from one phase of the cycle to another.
Interactions between the CPG neurons are flexible and can be modified by the same command inputs that activate the CPG. In mammals, the descending command inputs not only turn on the locomotor CPG, but also establish a mode of interaction between the elementary CPGs controlling rhythmic movements of different limbs. It has been noted that higher concentration of NMDA and serotonin decrease reciprocal right/left inhibition in the rat spinal cord. It could play a role in the transition from diagonal gaits to the gallop upon increasing stimulation of the mesencephalic locomotor region.
The final shape of the motor pattern is determined by synaptic connections between the generator neurons and the neurons of the output stage of the CPG, as well as synaptic connections within the output stage. The relatively simple output of the rhythm generator is transformed into a more complicated pattern of activity of the motor neurons due to the convergence of excitatory and inhibitory influences from the rhythm generator (see Figure 1G,H)
Although CPGs produce the basic pattern of a motor output, motoneurons are not passive followers in most cases. their properties can be modulated. For example, some spinal motoneurons in vertebrates can generate prolonged plateau potentials under the influence of monoaminergic inputs. As a result of plateau properties, the responses of motoneurons to synaptic inputs are amplified in intensity and duration.

4 Role of sensory feedback in producing and shaping motor pattern

The role of sensory feedback can be illustrated when considering locomotion. The control of locomotion in a homogeneous medium (water, air) requires less sensory information than the control of locomotion in irregular environment, for example, walking on the ground where in each step the leg can be affected by irregularities of the substrate. Movements performed by different species differs in the role that is played by central mechanisms (CPGs) and sensory feedback. At one extreme we have aquatic animals whose locomotor system needs minimal or no feedback for its function, and the final motor pattern is generated by the CPG. The locomotor systems controlling wing flapping in Clione, body undulations in Aplysia, Tritonia, leech, and tadpole belong to this class. Close to them are the locomotor systems controlling swimming body undulations in the lamprey and wing beating in the locust.

At the other extreme we find the species whose locomotor systems practically cannot operate under open-loop conditions. For example, the stick insect (which normally moves in extremely irregular environment, climbing on branches of the trees). Deprived of sensory feedback, movements of individual joints of a leg become uncoordinated, and inter-leg coordination also suffers.

Finally, the locomotor systems of the crayfish, lobster, locust and cat, performing locomotion on ground, occupy an intermediate position. Under open-loop condition, they can generate rhythmic activity more or less resembling a normal locomotor pattern. A closed feedback loop is necessary for their normal function, however. Under close-loop condition, the timing of different events in the step cycle is largely depended not on the CPG activity, but on the afferent signals about limb movement. In the cat, a critical role in determining transition from the stance to the swing phase of the step is played by the signals about hip position and about unloading of the limb (see Figure 4B-D). Several reflex mechanisms adapt the limb movements to external conditions. In the stance phase, the extensor activity is modulated largely by the stretch reflex. In the swing phase, external stimuli can evoke modifications of the motor patter of the limb transfer (see Figure 4E-G).

The CPGs for different movements are systems of high reliability, which is attributable to their redundant organization. The characteristics of a CPG crucial for its operation are determined by a number of complementary factors acting in concert. However, various factors are weighted differently in determining different aspects of CPG operation. Whereas rhythm generation is mainly based on the pacemaker properties of some CPG neurons, neuron interactions play an important role in sculpturing the final motor output. The redundant organization of CPGs not only guarantees their reliability, but also allows them to be very flexible systems. Modulatory inputs from higher centers can influence different mechanisms involved in pattern generation and, thus, regulate CPG operation in perfect relation to a behaviorally relevant context.

Figure Captions

Figure 1: The locomotor CPG of the marine mollusc Clione. A. Schematic drawing of Clione (a ventral view). B. Successive wing positions during a locomotor cycle (a frontal view): (1) the maximal ventral flexion,

(2) the movement on dorsal direction, (3) the maximal dorsal flexion, (4) the movement in ventral direction. C. Activity of wing motoneurons (MN) during locomotiom. The MN 1A is active during dorsal flexion of the wing (D-phase) in a swim cycle, while 2A is active during ventral flexion (V-phase). NW - activity in the wing nerve. D. The locomotor pattern can be recorded in in vitro preparation consisting of isolated pedal ganglia (PedG) connected by the pedal commissure (PedC). The activity of MNs and interneurons is recorded intracellularly by microelectrodes (ME). Activity of MNs is also recorded extracellularly from NW with a suction electrode (SE). E. Activity of NW during fictive locomotion. F. Activity of two generator interneurons (from group 7 and 8) during fictive swimming. Excitation of a neuron of one group is accompanied by the appearance of the IPSP in a neuron of the antagonistic group. G, H. Locomotor CPG of Clione (G) and schematic pattern of activity of various cell groups in a swim cycle (H). The locomotor rhythm is generated by two groups (half-centers) of generator neurons (7 and 8) with mutual inhibitory connections. These interneurons produce EPSPs and IPSPs in MNs of D-phase (groups 1 and 3) and in those of Vphase (groups 2,4,6,and 10). Electrical connections between neurons are shown by resistor symbols, excitatory and inhibitory synapses by white and black arrows correspondingly. I-K. Experiments with isolation of generator interneurons. Activity of group 7 interneuron before extraction (I1,J1) and after extraction (I2,J2) from ganglion. J3. The effect produced by injection of various direct currents in the isolated group 7 interneuron. K. A single action potential can be evoked on rebound after injection of a pulse of hyperpolarizing current. L. Contribution of the rebound property in of generator interneurons to rhythm generation. In the absence of rhythmic activity in the pedal ganglia, the CPG can be triggered by a single pulse of hyperpolarizing current injected into interneuron 7. The black and white arrows show the appearance of IPSPs in interneuron 7 and in the swim MN 2. Due to the rebound, each IPSP gives rise to one half-cycle of the locomotor rhythm.

Figure 2: Feeding CPG of the pond snail Planorbis. A. Schematic drawing of Planorbis (a lateral view). When contact the food, the radula performs rhythmic movements. Retracted position of the radula is shown in B and protracted one (when the radula scratches the food object) in C. D. Preparation consisting of the buccal mass and buccal ganglia (BG) is capable of rhythmic radula (RAD) movements. These movements are shown in E together with activity in two buccal nerves (n1 and n2). Quiescence (Q), protractor (P) and retractor (R) phases are indicated. The same efferent pattern can be generated in the isolated buccal ganglia (F). G. Schematic pattern of activity of various cell groups in a feeding cycle. H-N.

The main features of the firing pattern of generator interneurons persisted in isolated cells. Activity of group 1e interneuron before extraction (H) and after extraction (I-K) from ganglion. Activity of group 2 interneuron before extraction (L) and after extraction (M,N) from ganglion. I-K,M,N. The effects produced by injection of various direct currents in the isolated group 1e interneuron (I-K) and group 2 interneuron (M,N). O,P. Morphology of interneurons of 1e and 2 groups and their location in buccal ganglia. Q. The feeding CPG of Planorbis. Due to intrinsic properties of 1e cells, their activity gradually increases in the Q and P phases. Because of the mutual electrical connections within group 1, 1e-cells provide an excitatory drive to 1d-cells, which activate protractor motoneurons. The 1d-cells exert an excitatory action upon group 2 neurons. Having reached the threshold, type 2 cells generate a rectangular wave of depolarizing potential and inhibit the group 1 cells, thus terminating the P-phase and simultaneously activating retractor motoneurons. Designations as in Figure 1.

Figure 3: The locomotor CPG of lamprey. A. Schematic drawing of lamprey (view from above). B. The EMG activity during active swimming. C. NMDA-induced oscillations of membrane potential and the effects produced by injection of various direct currents in the pharmacologically isolated cell of spinal cord. The spinal cord was treated by TTX, blocking synaptic interactions. D-G. Coordination of unitary oscillators. D. Experimntal arrangement for separate manipulation with excitability of neurons in different parts of the spinal cord. A chamber with a piece of the spinal cord was separated into three pools (rostral, middle and caudal) perfused with NMDA solutions at different consentration, and the activity of the motor neurons was recorded from five ventral roots. E,F. The motor pattern generated by the spinal cord depends on the NMDA concentration in different pools (indicated in M for the rostral/middle/caudal pools). E. With equal concentration in all pools, the wave propagates caudally. F. With higher consentration in the caudal pool, the wave propagates rostrally. G. Schematic illustration of the idea of trailing oscillators. In the chain of three oscillators, the oscillator with a shorter cycle period automatically becomes the leading one. Designations as in Figure 1. H. The segmental locomotor CPG consists of two symmetrical (left and right) half-centres, each of which comprises 3 groups of interneurons. The excitatory interneurons (E) excite inhibitory commissural interneurons (I) that cross the midline and inhibit all classes of neurons on the contralateral side, including motoneurons (MN), the lateral interneurons (L), which inhibit I interneurons. I-J. The effect of elimination of mutual inhibitory connections between the right and left half-centers on cycle period. The cycle period generated by each isolated (by longitudinal splitting of the spinal cord) half-center is shorter (J) than in the presence of inhibitory connections in intact spinal cord (I).

Figure 4 Role of sensory feedback in control of locomotion in cat. A. Schematic drawing of cat (lateral view). B-D. A critical role in determining transition from the stance to the swing phase of locomotor cycle is played by afferent signals about the hip position and unloading of the limb. B. Entrainment of the fictive locomotor activity (evoked in immobilized cat) by sinusoidal hip movements. To monitor activity of the CPG, knee extensor and flexor nerves, which are active during the stance and swing phases of the cycle, respectively, were recorded. C. Stretch a hip flexor results in earlier termination of swing phase and initiation of stance phase of locomotor cycle. D. Electrical stimulation of the extensor group 1 afferents (signaling about the loading the limb) results in prolongation of the stance phase of the cycle. E-G. Reflex mechanisms adapt the locomotor movements to external conditions. In the swing phase, external stimuli evoke modifications of the motor pattern of the limb transfer. E. Successive limb positions during stepping in the spinal cat. In frame 2, the dorsum of the foot hits an obstacle (the black square). A signal from cutaneous afferents evokes the flexion reflex, which is incorporated into the swing phase of the locomotor cycle and brings the foot well above the obstacle (frame 4). F-G. The successive stick diagrams for the swing phase in normal step (F) and in disturbed step (G) are shown. The limb position at the moment of hiting is indicated by an arrow (S).

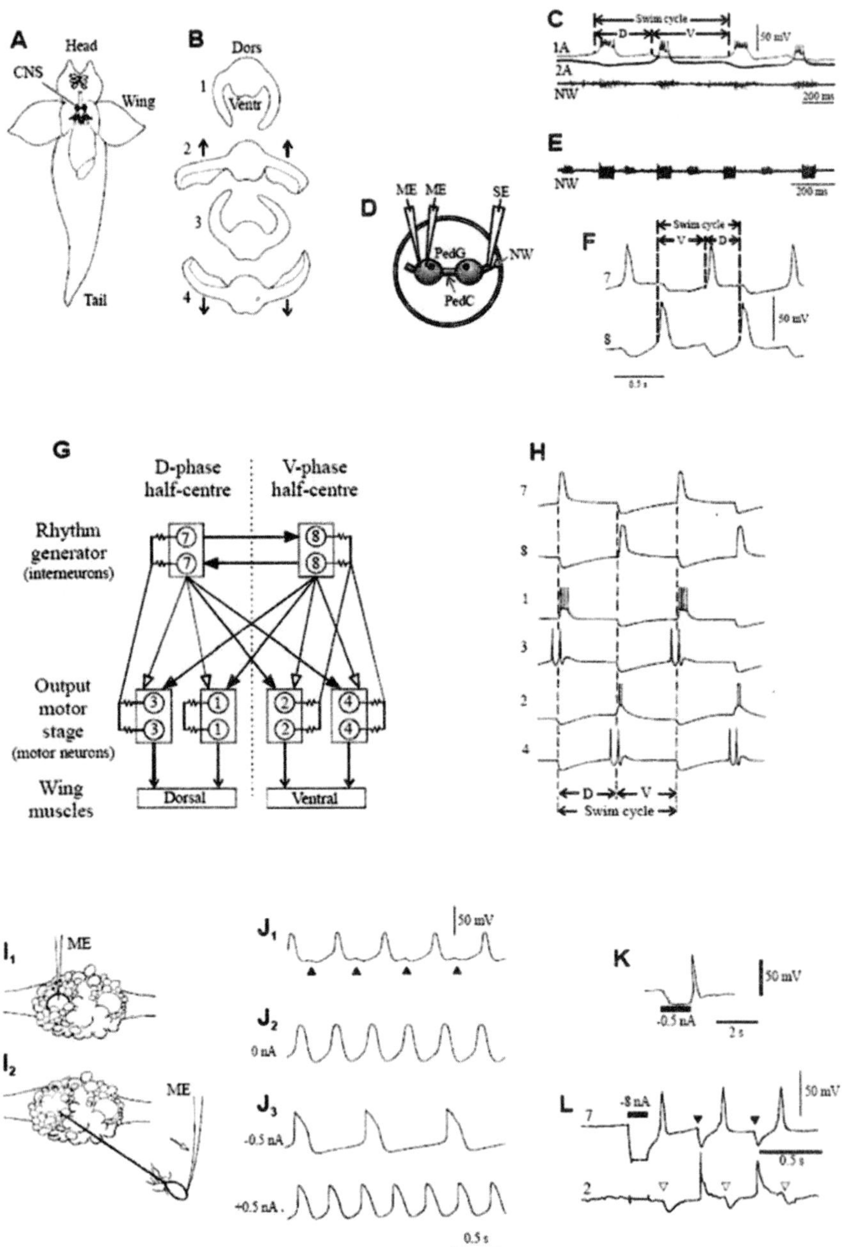

Figure 1. The locomotor CPG of the marine mollusc Clione.

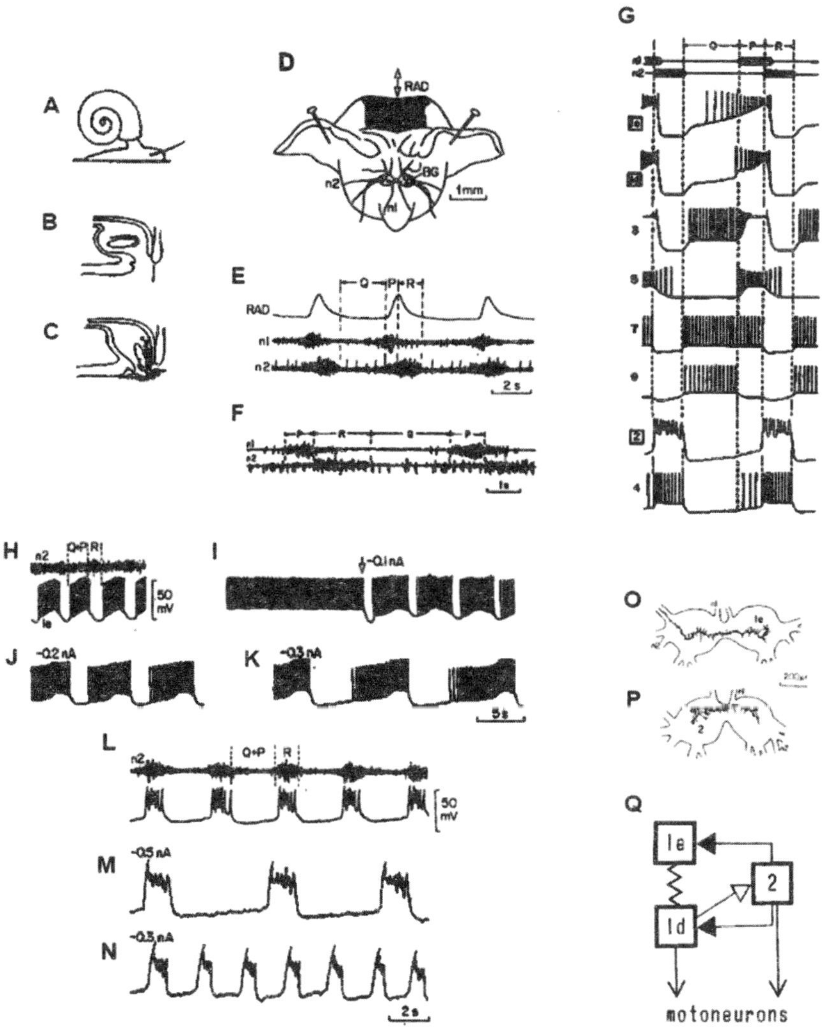

Figure 2. Feeding CPG of the pond snail Planorbis.

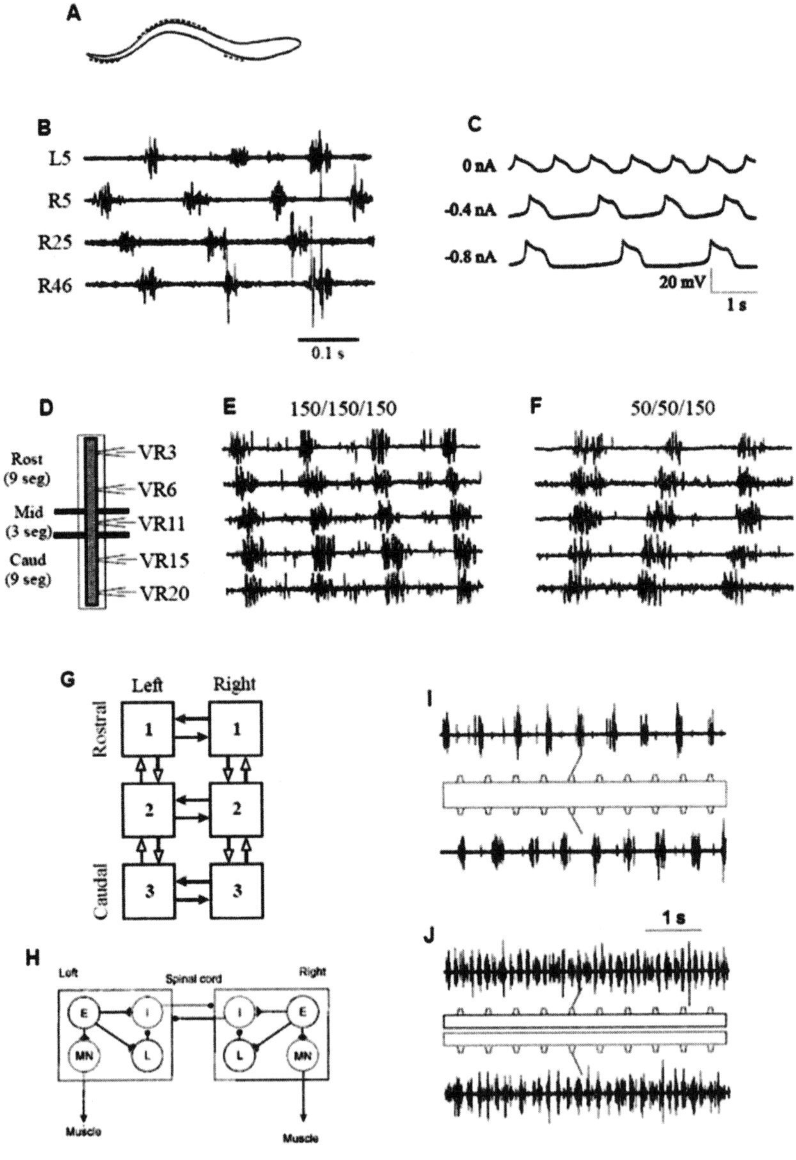

Figure 3. The locomotor CPG of lamprey.

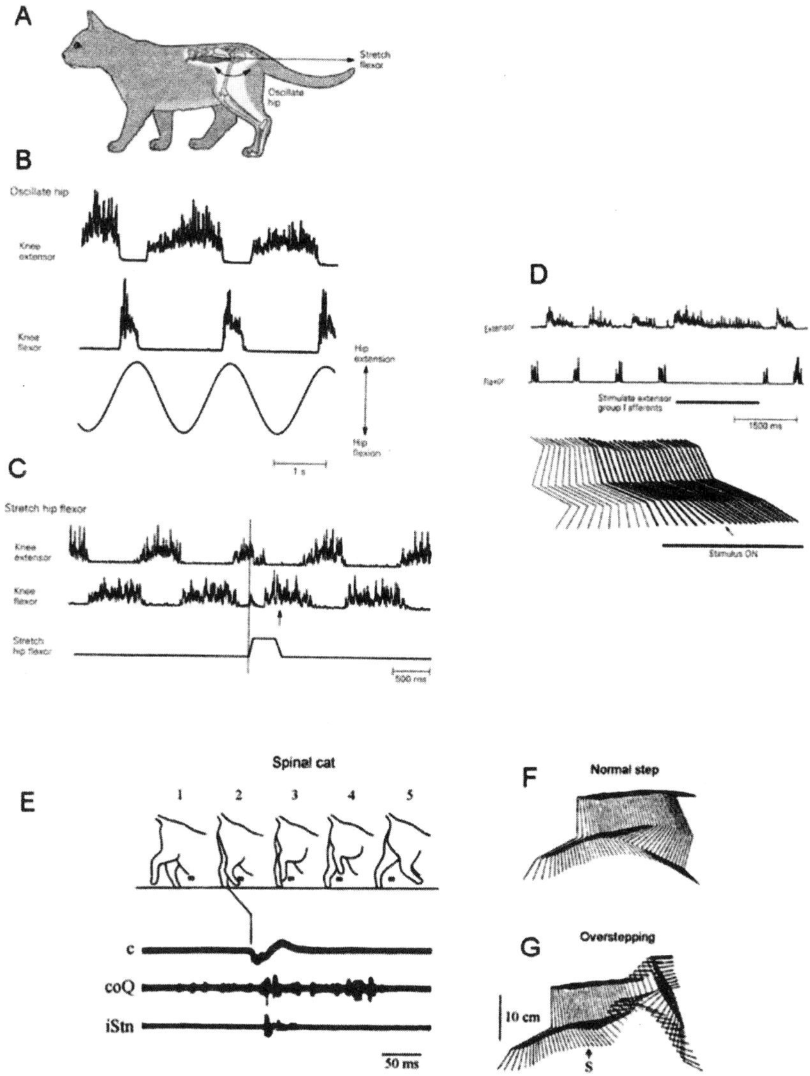

Figure 4. Role of sensory feedback in control of locomotion in cat.

Initiation and Generation of Movements: 2. Command Systems

Tatiana G. Deliagina
Department of Neuroscience, Karolinska Institute,
SE-171 77, Stockholm, Sweden

Abstract In both vertebrates and invertebrates, the CPGs for different movements can be activated by relatively simple (tonic) signals provided by command systems.

1 Command systems and command neurons in invertebrates

In the invertebrate animals, command systems for particular movements each contain relatively few neurons, and a contribution of individual neurons to activation of a CPG is substantial. Wiersma first demonstrated this in 1964. He found that electrical stimulation of a single neuron in the crayfish could evoke locomotor movements of swimmerets (locomotor organs), with the pattern very similar to that of normal swimming. In the subsequent studies it was shown that, in the CNS of crayfish and lobster, there are at least 5 pairs of neurons, each of which is capable of activating the locomotor CPG (see Figure 1). The neurons can be excited by the peripheral stimuli that evoke swimming. Later such neurons, which alone can activate the CPGs for different types of movement (locomotion, feeding, defence reactions, etc.) were found in various invertebrate species and termed the command neurons.

Detailed study of command neurons for swimming in the lobster revealed their important feature. Individual neurons are not functionally equivalent . the swimmeret beating they can elicit differs both in frequency range and in relative amplitude of strokes of different swimmerets. Individually, no single command neuron can elicit the full range of normal behavior. One can suggest that, under normal conditions, not a single but a group of command neurons is activated to elicit behavior. By activating differently different neurons, the animal can modify some features of the motor pattern generated by the CPG. Heterogeneity of command neurons was found also

in other command systems.
Both excitatory and inhibitory command neurons can participate in the control of a CPG, which makes the system more flexible. This is illustrated for the locomotor system of Clione. An excitatory command neuron Cr-SA, when activated by current injection, causes depolarisation of swim interneurons and activation of the locomotor CPG (see Figure 2A-C). An inhibitory effect of the command neuron I shown in Figure 2D for a neuron Pl-W2. Induced activity in Pl-W2 results in a complete inhibition of the locomotory rhythm.
Command neurons can affect not only the rhythm generator but also the output stage of the CPG, and even the muscles of the locomotor organs. In Clione, the CPC1 neuron, along with activation of the locomotor CPG (see Figure 4), enhances (through a special .modulatory. interneuron Pd-SW) contraction of wing muscles, which are elicited by signals from the CPG (see Figure 2E).
CCLsectionRole of command neurons in organization of complex behavior
Any form of complex behavior results from the coordinated activity of several motor centers. In Clione, locomotor activity is a necessary component of almost all forms of behavior, and this activity is coordinated with activities of other motor centers through a complex system of command neurons. We will consider two examples.

- Escape reaction. Mechanical stimulation of the tail during swimming dramatically increases the locomotor activity (frequency and amplitude of wing oscillations), and Clione tries to escape the irritant (see Figure 3A). This reaction is mediated by command neurons of the CPB1 group (see Figure 3B-E). The CPB1 neuron gets excited when tail mechanoreceptors are stimulated. In its turn, the CPB1 neuron (excited by current injection) strongly activates interneurons of the locomotor CPG and accelerates the locomotory rhythm. Along with activation of the locomotor system, the CPB1 neuron activates the heart excitatory neuron HE, which in its turn speeds up the heartbeat. Thus, the activities of two systems, the locomotor and circulatory ones, appear correlated.
- Hunting and feeding behavior. Clione is a predator; it feeds on a small mollusk Limacina. The hunting and feeding behavior is triggered by contact with Limacina (due to activation of mechano- and chemoreceptors), and has a number of components, including protraction of tentacles (to capture the prey), activation of locomotor CPG, and activation of feeding rhythm CPG (see Figure 4A). A pair of command neurons CPC1 plays a crucial role in the control of this complex behavior (see Figure 4B). They receive an excitatory sensory input signalling

about contact with the prey. These neurons exert widespread effects on different motor systems (see Figure 4C), including activation of the locomotor CPG, protraction of tentacles, and speeding up of the heartbeat. At the same time, the feeding rhythm CPG is activated by a different group of command neurons, PIN1. Thus, command neurons are responsible for coordination of different motor systems in complex forms of behavior. Activation of the locomotor CPG in different behavioural contexts can be performed by different command neurons, for example, by CPC1 during hunting (see Figure 4C), and by CPB1 during escape reaction (see Figure 3E).

2 Command systems in vertebrates

An evidence for the existence of command systems in vertebrates was first presented by Charles Sherrington about 100 years ago. Sherrington was studying the scratch reflex in cats and dogs. In response to irritation of the skin caused by, e.g., parasites, the animal protracts its hind limb toward the stimulated area. Upon reaching this area, the limb starts to rapidly oscillate (see Figure 5A,B). These rhythmic movements are aimed at removal of the irritant. They are generated by the CPG located in the lower spinal cord. A discovery by Sherrington was that the whole pattern of scratching could be evoked by electrical stimulation applied to a definite area in the upper spinal cord. Sherrington proposed a hypothesis that there exists a special group of neurons, which receive sensory input from cutaneous afferents innervating the receptive field of the scratch reflex. A tonic activity of these neurons, excited by sensory input, is transmitted by their axons down the spinal cord. These signals activate the spinal CPG for scratching (see Figure 5C). Later such groups of neurons, that integrate sensory inputs and activate the networks generating motor patterns, were termed the command systems.

The role of a command system is not only the activation of a CPG. As shown by Sherrington, the scratch reflex in the dog can be evoked from a very wide area of the skin (see Figure 5A). However, the hind limb performing scratching movements is always protracted towards the stimulated site. Thus, stimulation of more rostral area 2 causes larger protraction than stimulation of more caudal area 1 (Fig 5B). In other words, the generated motor pattern is somewhat different for different sites. This finding indicates that coordinates of the target for limb protraction are encoded in the signals transmitted by the neurons of command system. Sherrington suggested that these neurons constitute not a homogenous group but rather differ in their sensory inputs (see Figure 5C). Due to these differences, the population activity contains information about the stimulated site. Thus, the command system for scratching has a double function: first, it activates the spinal CPG and second, it determines some aspects of the motor pattern important for reaching the behavioral goal.

A powerful impulse for studying the command system for locomotion was given by the discovery of Shik and his colleagues. They showed in 1965 that electrical stimulation of a small area in the midbrain (mesencephalic locomotor region, MLR, see Figure 6A) could evoke coordinated locomotion in the decerebrate cat positioned on the moving belt of a treadmill (see Figure 6B,C). By simply increasing the strength of current pulses, one can force the animal to walk faster and to run, and even to switch from alternating limb movements to gallop. Later it was shown that MLR stimulation evokes locomotion also in intact cats. During locomotion, the cat passes by or jumps over the obstacles (see Figure 6D). This area (or analogous one) was found in different vertebrate species; its stimulation evokes walking in terrestrial quadrupeds including monkeys, flight in birds, and swimming in fish. Thus, the MLR is an essential part of the command system for locomotion in vertrbrates. How is this system organized?

In the cat and other mammals, signals from the brain to the spinal cord are transmitted through several descending pathways . the reticulospinal (RS) tract, the vestibulospinal tract, etc. A number of evidences suggest that, of these pathways, the RS tract is directly related to the initiation of locomotion, and specifically the RS neurons located in the two nuclei of the brain stem . nucleus reticularis gigantocellularis (NRGC) and nucleus reticularis magnocellularis (NRMC) (see Figure 6A): (1) These RS neurons receive excitatory input from the MLR. (2) They also receive excitatory input from the subthalamic locomotor region (SLR) . the other area whose stimulation also evokes locomotion (see Figure 6A). (3) Axons of RS neurons descend down the spinal cord and terminate in the areas were the locomotor CPGs reside. (4) Many RS neurons use an excitatory amino acid (glutamate) as a neurotransmitter, and application of glutamate or its agonists to the spinal cord promotes activation of the locomotor CPG. (5) Inactivation of the reticular nuclei by cooling reversibly blocks locomotion evoked by the MLR stimulation. (6) The locomotion evoked by continuous MLR stimulation or arising spontaneously is preceded and accompanied by excitation of RS neurons.

The scheme (see Figure 6D) shows functional organization of the command system for locomotion in the cat. Two locomotor areas (MLR and SLR) receive and integrate commands from the higher brain centers, which are responsible for the choice of behavior. The MLR and SLR represent two independent inputs to the reticulospinal system. These inputs are responsible

for a specific activation of those RS neurons, which constitute the principal part of the locomotor command system. They activate the spinal locomotor CPGs for each of the four limbs and regulate the intensity of locomotion.
The functional significance of double input to RS neurons is not clear. It was suggested that these two inputs are used for eliciting locomotion in different behavioural contexts, and for providing the locomotor pattern with some specific features. For instance, the appearance of the cat during locomotion evoked by SLR stimulation suggests that this locomotion may be associated with searching behavior.
A study of the command system for locomotion (swimming) in the lamprey revealed an important feature of this system that considerably simplifies the task of the control of locomotion by the CNS. Like in the cat, the command signals to the spinal CPG for swimming in the lamprey are transmitted by RS neurons. Different sensory inputs (visual, somatosensory, etc.) converge on RS neurons and can evoke swimming by activating these neurons. It was found that firing of RS neurons in intact lamprey could be maintained at a high level for a long period of time after termination of the stimulus that had elicited swimming. The swimming continues as long as the RS activity is high, for many seconds and even minutes (see Figure 7A). An explanation for this phenomenon is the specific membrane properties of RS neurons. In response to a brief stimulus, these neurons are able to generate long-lasting (plateau) potentials accompanied by continuous firing, which maintains the spinal locomotor CPG in an active state (see Figure 7B-D). Due to this cellular mechanism, a brief stimulus is transformed into a long-lasting motor response.
Another important feature of the command system for swimming is that a unilateral sensory input can initiate a bilateral, symmetrical activation of RS neurons (see Figure 7A), which is a necessary condition for rectilinear swimming. Thus, the command system is able to transform an asymmetrical stimulus into a symmetrical motor response. In other cases, when the animal wants to perform a turn, the commands transmitted by the left and right RS tracts occur different. This results in an asymmetrical activation of the left and right half-centers of the spinal locomotor CPG, and in turning.
In conclusion, a command system performs the following functions:

- It integrates sensory and central inputs related to the initiation of a given type of motor behavior.
- It activates a particular CPG or a group of CPGs necessary for generation of this behavior.
- It supplies the motor pattern with some specific features to adapt it to behavioural goals of the animal.

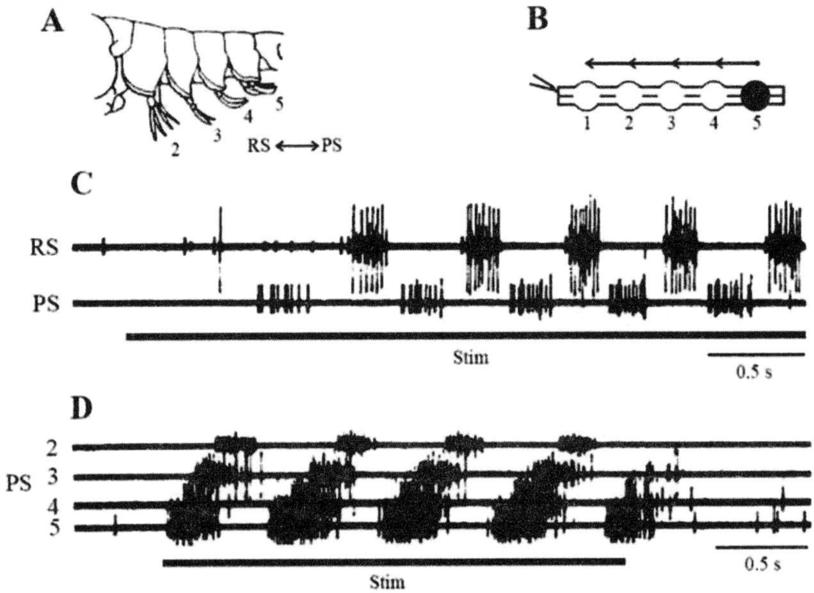

Figure 1. Initiation of swimming in lobster. A. Metachronal wave of beating in swimmerets 5-2 (PS -power stroke, RS - return stroke). B. Ganglia 2-5 of the nerve chord controlling swimmerets. The leading ganglion (pacemaker) is shown in black. An electrode stimulated the axon of the command neuron. C,D. Generation of a fictive swim pattern by an isolated chain of abdominal ganglia. C. Rhythmic motor output in the power stroke and return stroke nerves of a swimmeret, caused by stimulation (50 Hz) of a command neuron. D. Rhythmic motor output in the left power stroke nerves of ganglia 2-5, caused by stimulation of a command neuron.

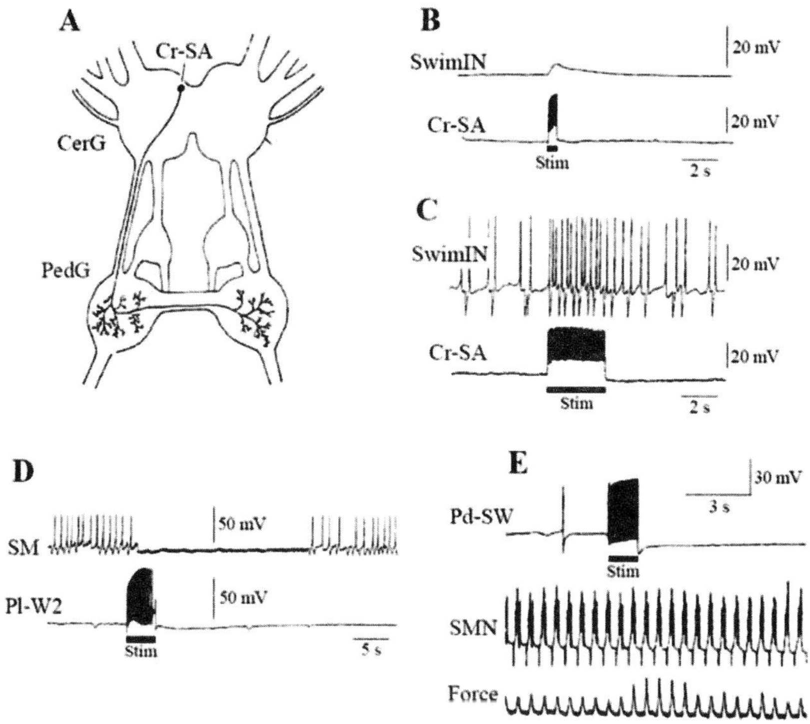

Figure 2. A-D. Command neurons produce excitation and inhibition of the locomotor CPG in Clione. A-C. The structure and action of a cerebral serotonergic anterior cell (Cr-SA). A. Structure. The cell has its cell body in the cerebral ganglion (CerG), and an axon that descends to and branches in both pedal ganglia (PedG). B. Discharge of Cr-SA (induced by current injection) elicits EPSP in the interneuron of the swim CPG (SwimIN). C. Induced repetitive firing of Cr-SA activates the locomotor CPG, as monitored by an increased frequency of SwimIN. D. The action of a pleural withdrawal cell (Pl-W2). Induced discharge of Pl-W2 evokes inhibition of locomotor activity. This is reflected in disappearance of rhythmical PSPs and spikes in a swim motor neuron (SM). E. Enhancement of contractivity of the wing muscles by serotonergic modulatory neuron (PD-SW). Discharges of a swim motor neuron (induced by periodical current injections) evoke contractions of the wing muscle. Activation of Pd-SW considerably increases the force of contraction.

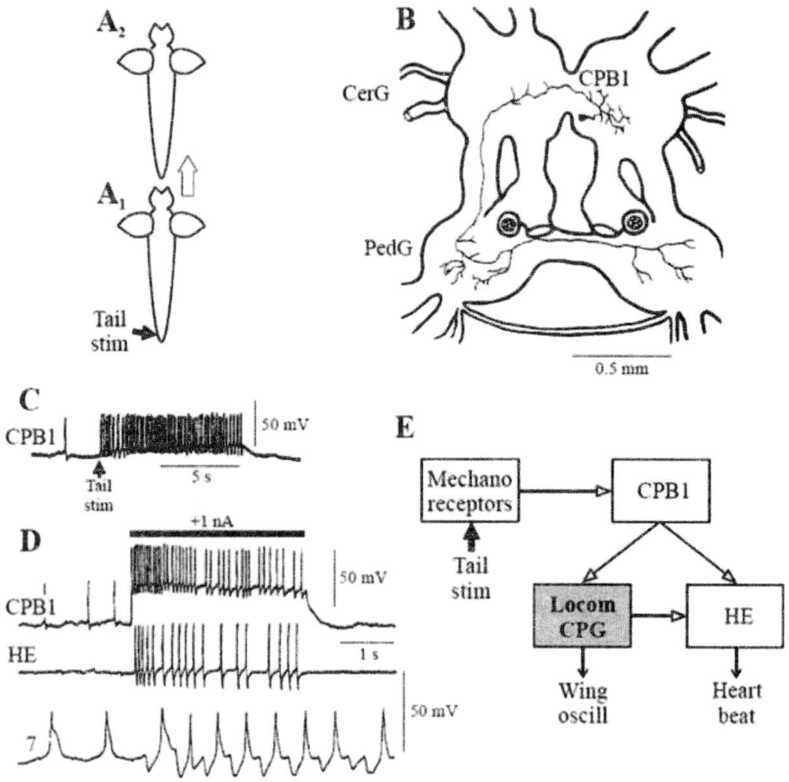

Figure 3. Formation of the synergy for avoidance reaction. A1,A2. Mechanical stimulation of the tail evokes fast swimming. B-E. The structure, input and output of the CPB1 neuron. B. Structure. It has a cell body in the cerebral ganglion (CerG), and an axon that descends to and branches in pedal ganglia (PedG). C. Input from the tail mechanoreceptors. D. Output to a swim CPG interneuron 7 and to the heart excitatory neuron HE, as revealed by activation of CPB1 evoked by current injection. E. Diagram of connections of CPB1. It receives excitatory input from the tail mechanoreceptors, and exerts an excitatory action on the locomotor CPG and on the heart excitor HE.

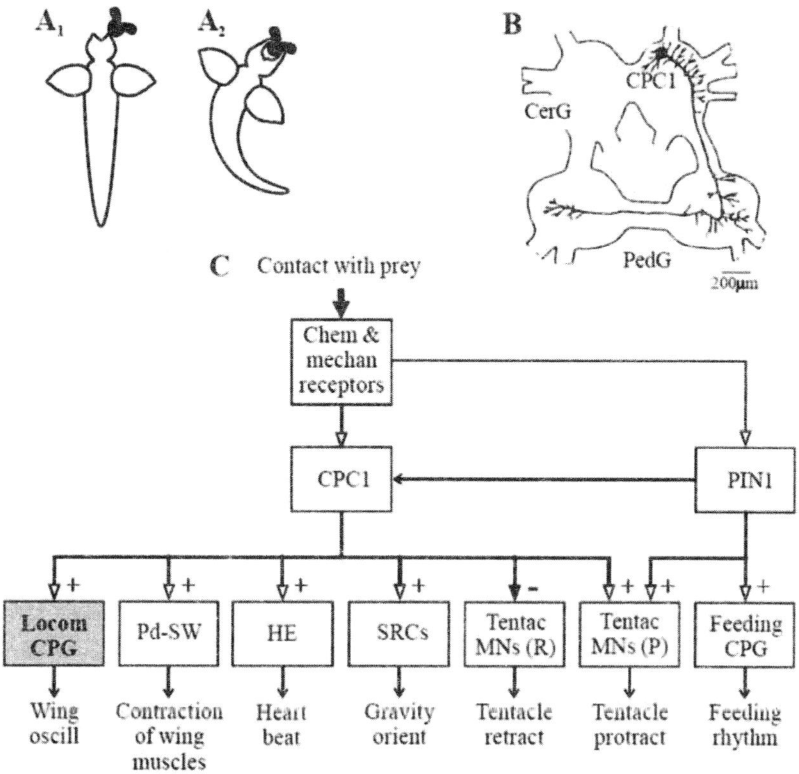

Figure 4. Formation of the synergy for hunting and feeding behavior. A1,A2. Contact with the prey (mollusc Limacina helicina) evokes protraction of head tentacles, turning towards the prey, acceleration of wing beating, and (when Limacina is captured) feeding movements of buccal apparatus. B,C. Structure, input and output of the CPC1 command neuron. B. Structure. It has a cell body in the cerebral ganglion (CerG) and an axon projecting to pedal ganglia (PedG). C. Diagram of connections of CPC1. The gross synergy for hunting and feeding is primarily formed due to the action of CPC1 upon different motor systems. However, PIN1 command neurons also contribute to the formation of this synergy. Targets of CPC1 and PIN1 are: the locomotor CPG, PD-SW modulatory neurons, heart excitor (HE), statocyst receptor cells (SRCs), tentacular protractor (P) and retractor motor neurons, and feeding CPG.

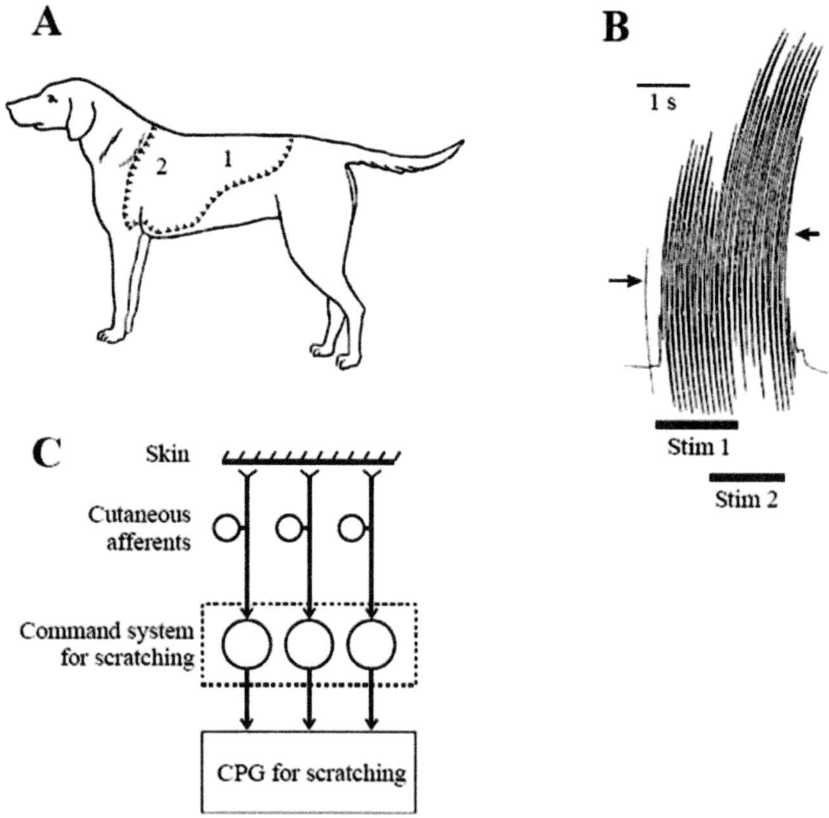

Figure 5. Scratch reflex in the dog. A. The spinal dog (transection in upper thoracic region) is able to scratch different sites within the receptive field (shown by broken line). B. The protraction-retraction movements of the hind limb evoked by stimulation of sites 1 and 2 of the receptive field. C. A command system for the scratch reflex is formed by propriospinal neurons activated by cutaneous afferents.

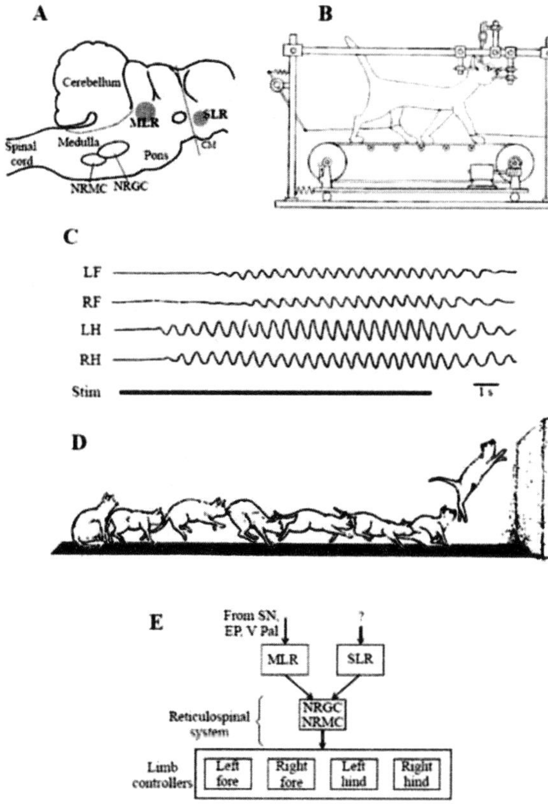

Figure 6. Command system for locomotion in the cat. A. Two areas in the brain stem related to initiation of locomotion: the mesencephalic locoomotor region (MLR) and subthalamic locomotor region (SLR). Their effects on the spinal locomotor CPGs are mediated by the nucleus reticularis gigantocellularis (NRGC) and nucleus reticularis magnocellularis (NRMC). The level of decerebration is shown (CM - corpus mammillare). B. Experimental arrangement to study evoked locomotion in the decerebrate cat. The cat is fixed in the stereotaxic device, with its four legs walking on the belt of treadmill. The MLR is stimulated (pulses 20-50 Hz). C. Locomotor episode evoked by MLR stimulation (LF - left fore limb, RF - right fore limb, LH - left hind limb, RH - right hind limb). D. Locomotor activity of the intact cat evoked by electrical stimulation of MLR. Stimulation (20-50 Hz) was performed through the chronically implanted electrodes. Time intervals between the successive video frames are 0.1 s. E. An overview of the structures involved in the initiation of locomotion. The MLR receives inputs from higher motor centers . the entopeduncular nucleus (EP), substantia nigra (SN), and ventral pallidum (V. Pal).

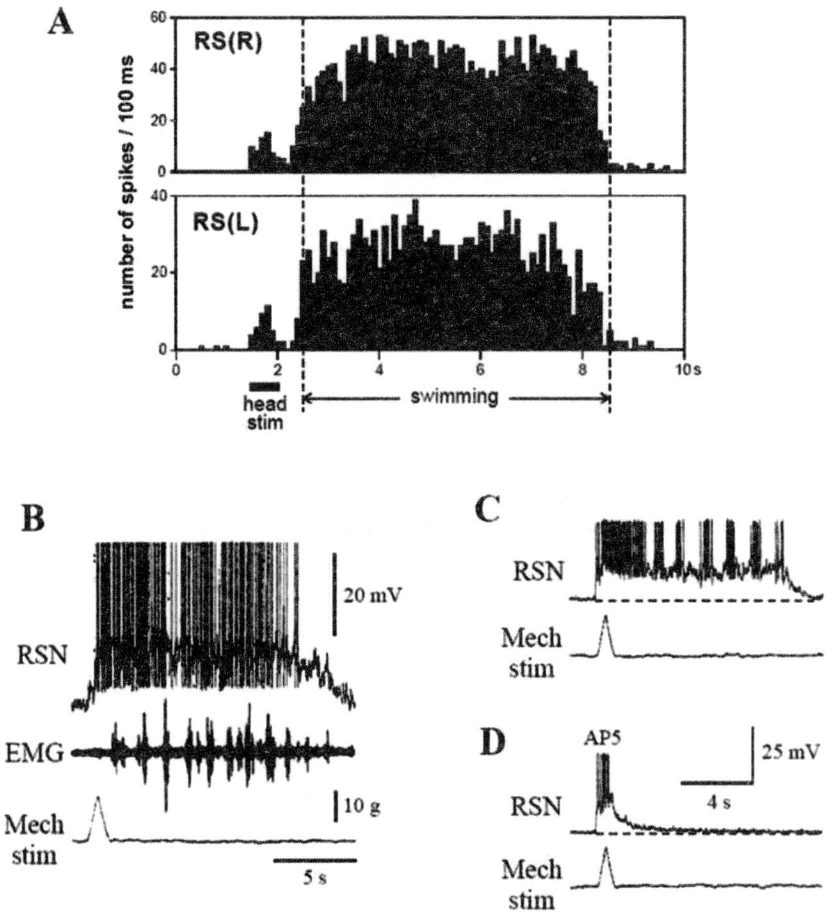

Figure 7. Initiation of swimming in the lamprey. A. Histograms of the population activity of the larger RS neurons recorded from the left (L) and right (R) sides of the spinal cord of the intact lamprey by means of implanted electrodes. Tactile stimulation of the head evoked strong bilateral activation of RS neurons and swimming. B-D. Effects of the mechanical stimulation of the skin covering the head region in the semi-intact preparation. B. Brief stimulus elicited a plateau potential and spike activity in the RS neuron accompanied by the rhythmic EMG bursting and undulatory movements of the caudal part of the body. C,D. The plateau potential in the RS neuron evoked by skin stimulation (C) was dramatically reduced by local application of AP5 (blocker of NMDA receptors) to the somata of the neuron (D).

Stabilization of Posture

Tatiana G. Deliagina
Department of Neuroscience, Karolinska Institute,
SE-171 77, Stockholm, Sweden

Abstract Different species, from mollusk to man, actively maintain a basic body posture (that is a particular orientation of their body in space) due to the activity of postural control system. For example, marine mollusk Clione and man maintain the vertical (headup orientation), the fish and terrestrial quadrupeds maintain the dorsal side-up body orientation. Deviations in any plane from this orientation evoke corrective movements, which lead to a restoration of the initial orientation. The stabile posture also presents a basis on which voluntary movements of different parts of the body can be superimposed. Maintenance of body posture is a non-volitional activity that is based, in many species, on innate neural mechanisms. Postural systems differ from those for movement control in their behavioral goals. The systems for movement control cause a movement of the whole body or its segments from one position in space to the other, as in walking or reaching. The systems for postural control prevent movements. they stabilize a position (or orientation) of the body in space, or orientation of its segments in space and in relation to each other.
Two principal modes of postural activity can be distinguished: (1) The feedback mode is compensation for the deviation from the desired posture (see Figure 1A). (2) The feedforward mode is anticipatory postural adjustments aimed at counteracting the destabilizing consequences of voluntary movements (see Figure 1B). In this lecture I will focus on the feedback mode of postural activity.

1 Functional organization of postural system

The extreme importance of postural control in humans stimulated numerous studies in this field. They led to a formulation of the hypothesis about functional organization of postural system that stabilizes body orientation (see Figure 1C). This closedloop system operates on the basis of sensory information about body orientation delivered by vestibular, visual, and somatosensory inputs. These signals are processed and integrated to obtain a general characteristic of the current body orientation (like position of its center of mass, or orientation of its axis in relation to the vertical). This characteristic is termed .a regulated variable. If the current value of the regulated variable differs from the desirable one, a corrective motor command is generated. The command elicits a motor response aimed at restoration of the initial orientation.

Studies in the field of postural control are devoted to different aspects of this general scheme. The most common method in these studies is observation of motor responses to postural disturbances. The main conclusions from these studies are the following:

1. Processing and integration of sensory inputs. The relative role of different sensory inputs in postural stabilization is species-dependent. In particular, vestibular input plays a much larger role in aquatic animals than in terrestrial ones (Figs. 1D and 2). A relative contribution of inputs of different modalities for a particular species is not constant but may vary considerably depending on the behavioral state of the subject (see Figure 1D) and environmental factors. For instance humans, when they are stabilizing the vertical body orientation, relay on somatosensory information if standing on the solid surface, and on vestibular information if standing on the soft surface.
2. Body configuration and equilibrium. In most species, a body consists of many segments, each of which must be stabilized in relation to other segments, as well as to the external coordinate system. It is suggested that the CNS subdivides this complex task into two simpler ones. maintenance of body configuration and maintenance of equilibrium. and solves them separately. For example, the cat can maintain the dorsal side-up trunk orientation at different configuration of its limbs (hemi-flexed, extended), and with different inter-limb distance.
3. Corrective motor responses. Disturbances of the upright body posture may differ in their direction, magnitude, etc. It is suggested that, to cope with these infinitely variable disturbances, a special strategy is used. This strategy includes a selection of the appropriate class of response (the muscle synergy) from a limited set of classes, and regulation of the value of the response (see Figure 1A).

These conclusions and concepts relate mainly to the functional organization of postural control. Much less is known, however, about the organization and operation of the corresponding neuronal networks. In particular, it is not known how and where in the CNS sensory inputs are processes and integrated to compute the regulated variable. Another important question is how and where the desirable value of the regulated variable is set, how and where the signals, which code the current and desirable values of the regulated variable, are compared, and how and where the commands for postural corrections are generated. These questions are difficult to address in higher vertebrates because of extreme complexity of their postural system that includes numerous sensory and motor centers interacting with each other. In contrast, .simple. animals present more opportunities for the analysis of postural networks.

2 Postural networks

Organization and operation of postural networks was studied in detail in two simple animal models . the invertebrate animal (mollusk) Clione and the lower vertebrate animal lamprey. Both animals are aquatic ones. They actively stabilize their orientation in the gravity field by using vestibular information. In both animals, some environmental factors cause a change of the stabilized orientation.

The postural control system of Clione is responsible for stabilization of body orientation in any vertical plane. For each particular plane, the system includes two chains of antagonistic tail reflexes driven by gravitational input from two statocysts (gravity sensitive organs). The system stabilizes the orientation at which the two reflexes compensate for each other (an equilibrium point of the system). Normally this occurs at the vertical, head-up orientation (see Figure 2B). Raising the water temperature causes a dramatic reconfiguration of the network, and a reversal of postural reflexes (see Figure 2A). This leads to a change of the equilibrium point in the system from the head-up orientation to the head-down orientation (see Figure 2C). As a result, the animal swims downward in an attempt to reach colder layers of water. The system is also able to change gradually the stabilized orientation by changing the gain in one of the reflex chains (see Figure 2D).

In the lamprey, the postural system can be subdivided into the roll and pitch control systems stabilizing body orientation in the transverse and sagittal planes, respectively. Operation of each system is based on the interactions between two antagonistic vestibular postural reflexes, mediated by two groups of reticulospinal (RS) neurons causing rotation of the animal in the opposite directions, as illustrated for the roll control system in Figure 2E. Due to vestibular input, activity of RS neurons depends on the orientation of the animal in the transverse plane (see Figure 2F). The system stabilizes the orientation at which the antagonistic reflexes compensate for each other (the equilibrium point). Normally, this occurs at the dorsal-side-up orientation. The stabilized orientation can be changed by asymmetrical eye illumination, which causes a shift of the equilibrium point (see Figure 2G), and a new orientation (with some roll tilt) will be stabilized.

The postural system is traditionally considered as the servo-system in which postural corrections are caused by the signals about deviation of the regulated variable (a body axis or position of the center of mass) from its desirable value (see Figure 1C). These signals cause a generation of postural corrections. However, studies on Clione and lamprey have shown that postural control can be based, at least in .simpler. animals, on a different principle, that is interaction between antagonistic postural reflexes. In both animals, stabilization of body orientation in a particular plane is based on the interaction of two antagonistic reflexes controlled by two groups of central neurons with opposite vestibular inputs. The system maintains the orientation at which the activities in the two groups are equal. In both animals, a stabilized orientation can be gradually regulated through a change of the gain in one of the reflex chains; this leads to a shift of the equilibrium point in the control system, and causes a dorsal light response in the lamprey and GABAinguced tilt in Clione. In addition to a gradual change of postural orientation, Clione is able to switch between the two distinct postural orientations, head-up at lower temperature and head-down at higher temperature, which is due to a reconfiguration of the postural network.
The similarity in operation of postural systems revealed in two evolutionary remote species support the hypothesis that such a basic problem as the neuronal control of antigravity behavior has a similar solution in different species, and that principles revealed in simpler animal models may have more general significance and operate in higher vertebrates as well.

3 Sub-systems of postural system

The postural system in quadrupeds normally operates as a functional unit and stabilizes both the head and the trunk orientation (see Figure 3A,B). Under certain conditions, however, the system clearly dissociates into the sub-systems that independently control the head and the trunk. For example, the animal can stabilize the dorsal side-up orientation of its trunk on the tilting platform but, at the same time, it does not stabilize the head orientation (to perform movements by head), or it stabilizes the head orientation differing from that of the trunk.
Recent studies on the rabbit strongly suggest that lateral stability of the anterior and posterior body parts of quadrupeds are also maintained by two relatively independent sub-systems (see Figure 3D-J). Such a functional organization is similar to that of the locomotor system in quadrupeds, where the shoulder and hip girdles have their own control mechanisms, and even individual limbs have relatively autonomous controllers that generate stepping movements and interact with each other to secure inter-limb coordina-

tion. It seems likely that a control system consisting of semi-autonomous subsystems better adapts to complicated environmental conditions.
The sub-system responsible for stabilization of the head orientation is driven mainly by vestibular and visual inputs (see Figure 3C). By contrast, the sub-systems responsible for stabilization of the anterior and posterior parts of the trunk are driven by their own somatosensory inputs from corresponding limbs (see Figure 3C-J). It seems that receptors supplying postural networks with information about limb loading play an important role. For example, the signals from Golgi tendon organs might give rise to the "reversed" 1b load-compensating reflexes and thus promote postural stabilization.
It was hypothesized that each of the three postural sub-systems in quadrupeds (see Figure 3K) operates by using principle similar to that revealed in simpler animals models - that is, interaction of antagonistic postural reflexes. The sub-system stabilizes such orientation of a corresponding part of the body at which the effects of antagonistic postural reflexes are equal. To support or reject this hypothesis future experiments are necessary.

4 Localization of postural functions in mammalian CNS

Earlier studies have shown that chronic decerebrate animals (in which the brain was transected between the brainstem and forebrain) can sit, stand, and walk; when positioned on its side, the animal exhibits a set of righting reflexes and rapidly assumes the normal, dorsal-side-up posture. These findings indicate that an essential part of the nervous mechanisms responsible for the control of basic posture is located below the decerebration level, that is in the brainstem, cerebellum, and spinal cord.
The spinal cord plays a double role in the control of basic posture: it represents an "output stage" for supraspinal commands; and, owing to the spinal mechanisms activated by supraspinal drive and responding to .local. sensory inputs, it is directly involved in the generation of corrective postural responses. The relative importance of these two functions of the spinal cord is not clear. It is well established, however, that the animals with a complete transection of the spinal cord in a lower thoracic region exhibit poor postural responses and, as a rule, are not able to maintain the dorsal-side-up orientation of their hindquarters, though a reduced postural control may remain and can be improved by training.
These results may have two different interpretations. First, they can be considered as evidence for the minor role that is played by spinal postural reflexes for maintenance of body posture. The second, alternative hypothesis is that the transection of the spinal cord deprives the spinal postural

networks of the necessary supraspinal tonic drive, which results in a reduction in the activity of spinal postural reflex mechanisms. Indirect evidence for this hypothesis was obtained in lesion experiments. Animals subjected to a lateral hemisection of the spinal cord were, after some period of recovery, able to maintain equilibrium during locomotion and standing. As the hemisection causes dramatic changes both in ascending signals and in descending commands, one can suggest that the persistence of postural control after this lesion is due to the activity of spinal postural mechanisms. Even stronger evidence for this is the restoration of lateral stability in the hind quarters observed after bilateral hemisections. A recovery of postural muscle tone in spinal animals subjected to special training also supports this hypothesis.

The involvement of the brainstem and cerebellum in postural control has been confirmed in two lines of experiments. First, it was found that electrical stimulation of specific sites in the brain stem (dorsal and ventral tegmental field) and in the cerebellum (hook bundle) strongly affected the extensor muscle tone. These effects are mediated by reticulospinal and vestibulospinal pathways. Second, single neuron recordings in the intact cat walking on the tilted treadmill demonstrated that brainstem neurons (giving rise to descending tracts, vestibulospinal and reticulospinal), strongly changed their activity with a change of tilt angle. It remains unclear, however, if the tilt-related activity of these neurons is responsible for the generation of postural corrections, or only for modulation of postural responses generated by the spinal mechanisms. In humans, it has been suggested that the cerebellum is not involved in the initiation of postural corrections, but rather in "scaling" of corrective motor responses.

Until recently, participation of the forebrain in postural control was hypothesized mainly on the basis of clinical and lesion studies. It was shown recently that basal ganglia participate in regulation of muscle tone by affecting the level of activity of neurons in the "ventral tegmental field". Participation of the motor cortex in the control of basic posture was directly demonstrated in recent experiments by recording activity of the motor cortex neurons in the rabbit during postural corrections. Most neurons of corticospinal tract were modulated during postural corrections caused by the lateral tilts of the supporting platform. Functional significance of these corticofugal signals is not clear, however, since integrity of the motor cortex is not necessary for stabilization of the basic body posture.

5 Conclusions

The following conclusions can be drawn:

- The basic body posture (upright in humans and dorsal side up in quadrupeds) is maintained by the closed-loop control system driven by sensory inputs of different modalities.
- There are two hypotheses concerning functional organization of the system . servo-control and reflex interactions.
- In quadrupeds, the postural system consists of three relatively autonomous sub-systems stabilizing positions of the head, and anterior and posterior trunk.

Figure Captions

Figure1 A,B. Two principal modes of postural activity: the feedback mode (compensation for the deviation from the desired posture, A1,A2) and the feed-forward mode (anticipatory postural adjustments aimed at counteracting the destabilizing consequences of voluntary movements, B1,B2). Backward (A1) or forward (A2) movement of the platform makes the subject sway forward or backward, respectively. This elicits a corrective motor response. Forward body sway (A1) evokes activation of extensor muscles (Postural muscle synergy 1). Backward body sway (A2) evokes activation of flexor muscles (Postural muscle synergy 2). B1,B2. The subject stands in a firm platform and pulls on a fixed handle as soon as possible after an auditory stimulus (arrow). To maintain posture, backward-acting contraction of the leg muscle (gastrocnemius) starts before the biceps begin pulling the handle. In A,B traces are electromiograms (rectified and integrated) of: Para, paraspinal; Abd, abdominal; Ham, hamstring; Quad, quadriceps; Gast, gastrocnemius; Tib, tibialis anterior muscles. C. General functional organization of postural control system. Sensory signals are processed and integrated to characterize the current body position. If it differs from the desirable one, a corrective motor command is generated. The command elicits a motor response aimed at restoration of the desirable orientation. D. One type of sensory information is sufficient for ability to stabilize the vertical body orientation. Mean sway over 20 s of stance is shown under six different sensory conditions in normal subjects and in patients with vestibular loss, and a sensory organization of deficit (Sensory conditions, bottom). Anterior/posterior peak-to-peak sway at the hips is normalized for each subject's height such that 100fall. Sensory conditions include blindfolding (conditions 2 and 5), sway-referencing the visual surround (conditions 3 and 6), and sway-referencing the support surface (conditions 4-6).

Figure2: Postural stabilization in Clione and lamprey is based on antagonistic gravitational reflexes. A. Postural network in Clione, responsible for stabilization of body orientation in the frontal plane, is driven by statocyst receptor cells (SRC) sensitive to the left (L) or the right (R) tilt. The SRCs, through two groups of CPB3 interneurons, excite tail motoneurons flexing the tail to the left, TMN(L), or to the right, TMN(R). Functioning of the network depends on which set of SRC-CPB3 connections (1 or 2) is activated. At low water temperature (10^oC), connections 1 operate, and the network stabilizes the vertical, head-up orientation. This is shown in B where the activities of TMN(L) and TMN(R) are plotted against the tilt angle. The arrows indicate the directions of rotation caused by the corresponding motoneurons. The activities are tilt-dependent due to the inputs from SRCs; they are equal to each other at 0° (an equilibrium point of the system). At higher water temperature (20^oC), connections 2 operate, and the network stabilizes the vertical, head-down orientation due to reversal of gravitational reflexes (C). In the intact animal, a gradual change of the stabilized orientation can be caused by GABA injection In the isolated CNS, GABA selectively reduces the gain in the right reflex chain, which results in a shift of the equilibrium point, and will lead to the tilt of Clione to the left (D). E. A conceptual model of the postural system responsible for stabilization of the dorsal-side up orientation in the lamprey. The key elements of the model are the left and right groups of reticulospinal neurons, RS(L) and RS(R). They receive vestibular (V) and visual (E) inputs, and through spinal mechanisms, evoke corrective postural responses, that is rolling to the left or to the right. Normally, without visual input, the system stabilizes the dorsal-side-up orientation. This is shown in F where the activities of RS(L) and RS(R) are plotted against the tilt angle. The arrows indicate the directions of rotation caused by the corresponding groups of RS neurons. The activities are tilt-dependent due to vestibular inputs; they are equal to each other at 0o (an equilibrium point of the system). A gradual change of the stabilized orientation can be caused by eye illumination. The illumination of the left eye increases the gain in the left reflex chain, and reduces the gain in the right reflex chain (G), which results in a shift of the equilibrium point of the system, and in the tilt to the left.

Figure3:A,B. Experimental design for testing postural corrections in rabbit. The animal was standing on a platform (A,D). A tilt of the platform caused an extension of the limbs on the side moving down and flexion on the opposite side (B). These limb movements made the trunk move in the transverse plane in relation to the platform, in a direction opposite to the platform tilt. This compensatory trunk movement reduced a deviation of the body from the dorsal-side-up position. Simultaneously the corrective movement of the head which brings the dorso-ventral axis of the head toward the vertical is observed (B). C. Absence of visual and vestibular information does not affect stabilization of the trunk and completely abolishes stabilization of the head. A coefficient of postural stabilization is shown for head and trunk before (Control) and after bilateral labyrinthectomy (BL) during visual deprivation (eyes closed). D-J. Corrective responses to complex perturbations of posture. The rabbit was positioned on two platforms, one for the fore limbs and one for the hind limbs, subjected to periodical lateral tilts (P1 and P2 in F). The platforms could be tilted either simultaneously, or in anti-phase, or at different frequencies. Postural corrections, that is lateral displacements of the anterior and posterior body parts in relation to the platforms, were recorded by mechanical sensors (S1 and S2 in D-F). With in-phase tilts of the platforms, the animal stabilized its dorsal-sideup position by displacing the whole body in the direction opposite to tilt (E,G). These compensatory body movements were caused by simultaneous extension of the fore and hind limbs on the side moving down, and flexion of the opposite limbs. With anti-phase tilts of the two platforms, the animal also maintained its dorsal-side-up position, though in this case the compensatory movements of the anterior and posterior body parts were in anti-phase (J); they were caused by the anti-phase flexion/extension movements of the ipsilateral fore and hind limbs. The rabbit was also able to stabilize its dorsal-side-up orientation when the platforms were tilted at different frequencies, or when one platform was tilted and the other was not (H,I). These data suggest that the anterior and posterior parts of the body have separate postural control mechanisms driven by their own somatosensory inputs. K. Organization of postural control system in quadrupeds. The system consists of three sub-systems responsible for stabilization of the head, anterior and posterior parts of the trunk, respectively. The subsystem responsible for stabilization of the head is driven by vestibular and visual inputs. By contrast, the sub-systems responsible for stabilization of the anterior and posterior part of the trunk are driven by their own somatosensory inputs from corresponding limbs.

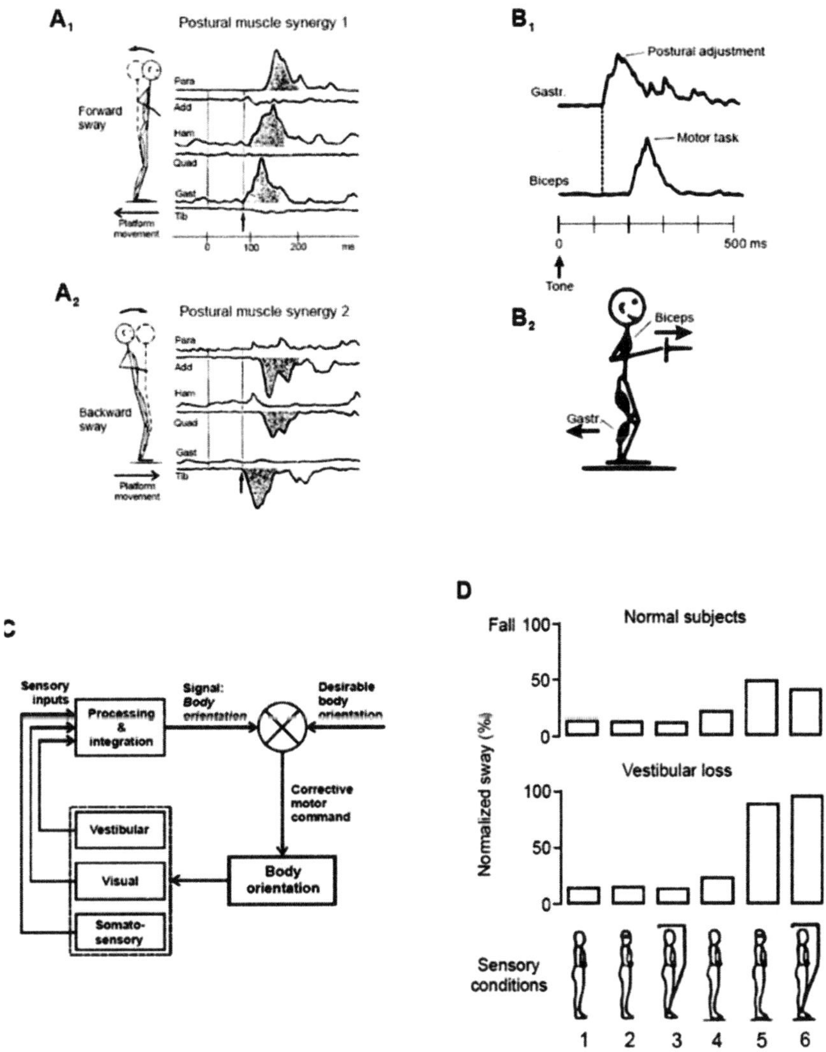
A1
Postural muscle synergy 1
Forward sway
Platform movement
Para
Add
Ham
Quad
Gast
Tib
0
100
200
ms
A2
Postural muscle synergy 2
Backward sway
Platform movement
B1
Postural adjustment
Gastr.
Motor task
Biceps
0
500 ms
Tone
B2
Biceps
Gastr.
C
Sensory inputs
Processing & integration
Signal: Body orientation
Desirable body orientation
Corrective motor command
Vestibular
Visual
Somato-sensory
Body orientation
D
Fall 100
Normal subjects
50
0
Vestibular loss
100
50
0
Normalized sway (%)
Sensory conditions
1
2
3
4
5
6

Figure 1.

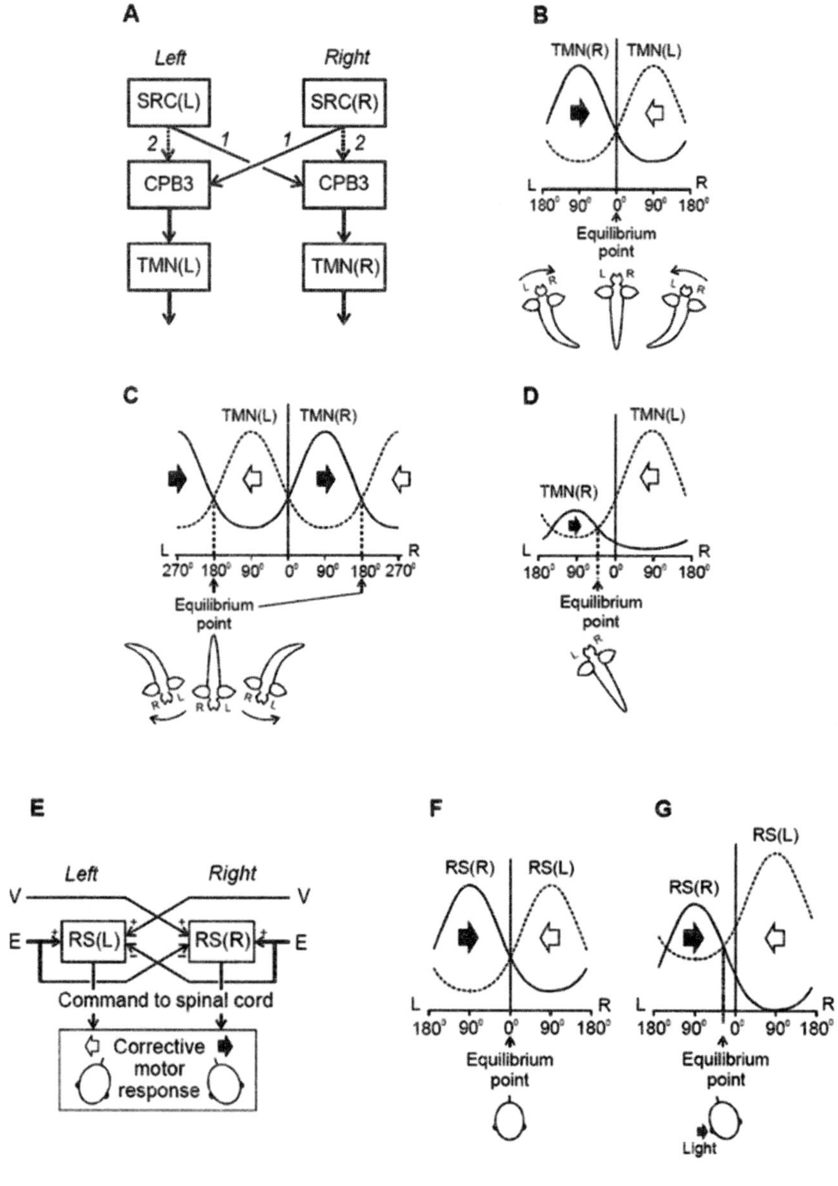
A
Left
Right
SRC(L)
SRC(R)
CPB3
CPB3
TMN(L)
TMN(R)
B
TMN(R)
TMN(L)
Equilibrium point
C
TMN(L)
TMN(R)
Equilibrium point
D
TMN(L)
TMN(R)
Equilibrium point
E
Left
Right
RS(L)
RS(R)
Command to spinal cord
Corrective motor response
F
RS(R)
RS(L)
Equilibrium point
G
RS(L)
RS(R)
Equilibrium point
Light

Figure 2.

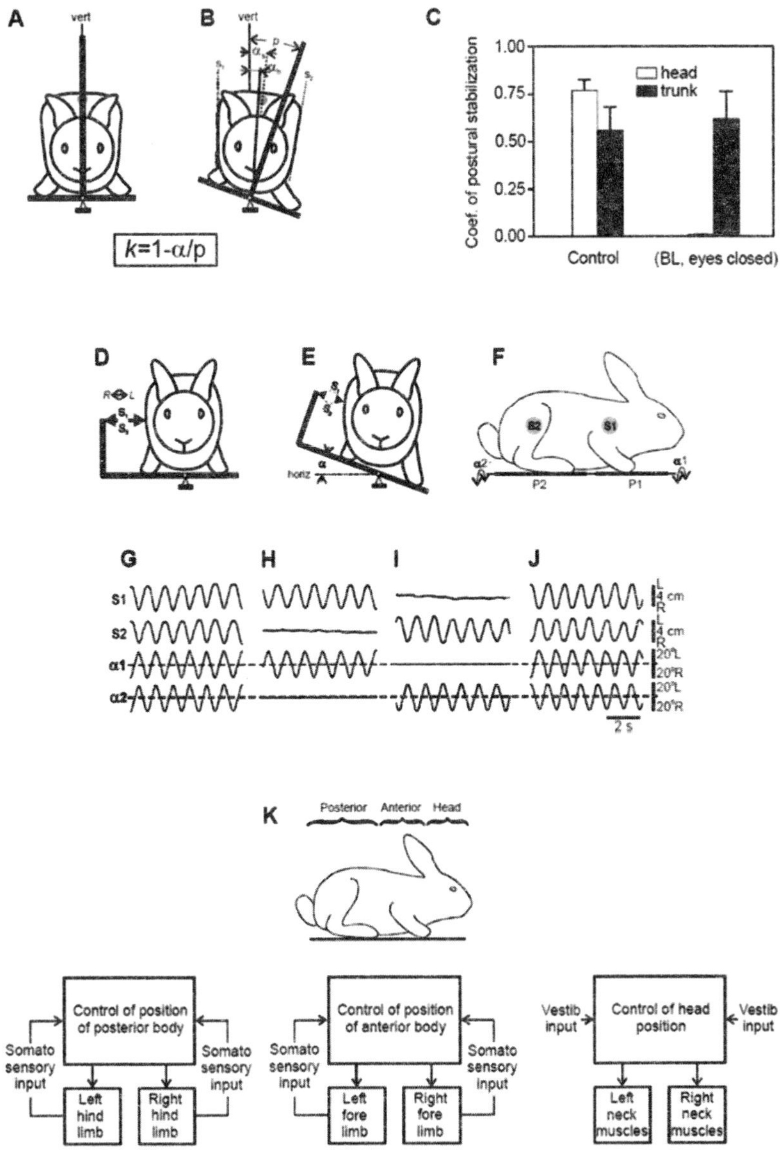
A
vert
B
vert
k=1-α/p
C
Coef. of postural stabilization
1.00
0.75
0.50
0.25
0.00
head
trunk
Control
(BL, eyes closed)
D
E
F
S2
S1
α2
α1
P2
P1
G
H
I
J
S1
S2
α1
α2
4 cm
4 cm
20°L
20°R
20°L
20°R
2 s
K
Posterior
Anterior
Head
Control of position of posterior body
Somato sensory input
Somato sensory input
Left hind limb
Right hind limb
Control of position of anterior body
Somato sensory input
Somato sensory input
Left fore limb
Right fore limb
Vestib input
Control of head position
Vestib input
Left neck muscles
Right neck muscles

Figure 3.

Locomotion as a Spatial-temporal Phenomenon: Models of the Central Pattern Generator

Paolo Arena

Dipartimento di Ingegneria Elettrica Elettronica e dei Sistemi,
Universitá degli Studi di Catania, Catania, Italy

Abstract The development of new approaches and new architectures for locomotion control in legged robots is of high interest in the area of robotic and intelligent motion systems, especially when the solution is easy both to conceive and to implement.
This first lecture emphasizes analog neural processing structures to realize artificial locomotion in mechatronic devices. The main inspiration comes from the biological paradigm of the *Central Pattern Generator* (CPG), used to model the neural populations responsible for locomotion planning and control in animals. The approach presented here starts by considering locomotion by legs as a complex spatio-temporal non linear dynamical system, modelled referring to particular types of reaction-diffusion non linear partial differential equations. In the following lecture these Spatio-temporal phenomena are obtained implementing the whole mathematical model on a new Reaction-Diffusion Cellular Neural Network (RD-CNN) architecture. Wave-like solutions as well as patterns are obtained, able to induce and control locomotion in some prototypes of biologically inspired walking machines. The design of the CNN structure is subsequently realized by analog circuits; this gives the possibility to generate locomotion in real time and also to control the transition among several types of locomotion. The methodology presented is applied referring to the experimental prototype of an hexapod robot. In the last lecture the same approach will be shown to be able to realize locomotion generation and control in a number of different robotic structures, such as ring worm-like robots or lamprey-like robots.

1 Robotics and biologically inspired locomotion

Several walking and climbing machines were designed and developed during the last decades. Some of them were built to perform services for human utility and security (see for example CLAWAR (1998); Berns). In fact, among the various motion types, while wheels are nowadays still the most used way to realize motion in robots for their clear advantages with respect to the load carrying efficiency, legs are more attractive than wheels because they allow to reach places where only humans or animals on foot can go. This requires great adaptability to uneven or dangerous terrains. The problem of coordinating and controlling legs is also challenging and often researchers are helped from neurobiological studies on animal neuromotor systems, since even the simplest animals are able to move on legs and balance in a way that nowadays is challenging for artificial machines. Many legged robots were recently built, paying particular attention to joint motion and leg coordination and, under this point of view, a number of different control approaches were investigated. Some walking robots have a single pre-defined gait, while some others possess a number of fixed predetermined gait patterns and are able to switch between them. As regards the different approaches, those ones from Brooks (1999), from et al (1994) and from Ayers et al. (2000) are among the most important.

Brooks approach is directly focused to the control of multiple goals through the introduction of different levels of competence. Each level defines a particular class of behaviors: the upper levels define constraints on the lower levels, while each level can be dedicated to solve a particular task and is implemented via a finite state machine in a single processor, working asynchronously with the other ones, monitoring its own inputs and exchanging messages with the other processors. Once designed a control scheme, its implementation is realized via software or with digital microprocessors.

Pfeiffer's approach is based on the biological case of the stick insect, deeply studied by Dean et al. (1999). The hexapod robot, called TUM (et al, 1994), is based on two modules, a single leg controller (SLC) for each leg, and a leg coordination module (LCM). The SLC has the task to move the leg and to control its various phases, related to the particular position of the leg itself: the Anterior Extreme Position (AEP) and the Posterior Extreme Position (PEP). The LCM module, from information on the leg position, sets the AEP and the PEP for each leg, thus controlling the global behavior of the walking process. LCM computation aims to inhibit or excite each of the various phases (stance, swing) of a particular leg, depending on the phases in which the adjacent legs are. The model is based on some biologically based interleg coordinating rules and is implemented via software. Some other approaches were realized by using artificial neural networks simulators (Cruse et al., 1998).

Another biomimetic approach was performed by Ayers et al. (2000).He built an eight three-degree-of-freedom walking legs robot, mimicking the ameri-

can lobster *Homarus Americanus*. The neural based controller implemented by Ayers closely mimics electromyographic recordings of real walking in the lobster. The time signals reproduced from experimental readings are implemented using finite state machines algorithms. All of the above mentioned approaches, as well as almost all of the current approaches to biologically inspired robotics, have a more or less deep insight into the peculiarities of the motor generation and control in an animal. These approaches are prone to stop at an algorithmic stage, or they are implemented in a digital machine. In other words, traditional neuro-control systems take no consideration of the hardware in which they will be implemented. Moreover an efficient approach should require an efficient hardware implementation. Here a new approach to the real time generation and control of locomotion patterns is presented. The methodology takes into consideration the biological aspects of walking multipods, but never discards the implementation issues. The basic consideration is that living moving structures are constituted of a great number of degrees of freedom, concurrently actuated for a surprisingly efficient real time control. Under this perspective any digital approach is devoted to fail and the only framework is the analog circuit implementation. So the task is to design spatially distributed analog structures to work as neural pattern generators, able to manage in real time a great number of degrees of freedom, like biological neural tissues produce massively parallel signals to drive the muscular system. The methodology introduced takes its inspiration from the biological paradigm of the CPG, able to functionally model neural structures devoted to generate and control animal locomotion. Being the CPG a space distributed neural structure, a key point of the work was to design a suitable basic dynamics for an analog system, able to produce signals that qualitatively match neural dynamics. By direct comparison between neural signal properties and some models of spatio temporal dynamics arising in non linear active media it derives that signals propagating through living neurons belong to the class of autowaves (Krinsky, 1984). Thus the basic target was to design a spatio temporal analog circuit able to generate autowave fronts. The architecture built belongs to the class of CNNs. Subsequently a CPG structure based on CNNs was designed to generate suitable signals for locomotion generation and control purposes in biological inspired multipods.
The CNN approach to the implementation of the motion control has the peculiarities of the biological inspiration and the advantage of a low cost realization by means of analog circuits. It focuses on the realization of the locomotion task as a result of a robust oscillatory spatiotemporal dynamics (autowave) of the same type as neural firings. Locomotion is no longer the result of a high level approach implemented into a series of digital com-

mands on a digital processor, that could not manage with a true biological inspired machine made up of a great number of actuators and sensors, but it is a spatio temporal analog flow, the propagation of an analog wave. In particular this approach, reflecting the structure of the CPG, is divided into basically two different levels. The lower level realizes locomotion through an autowave propagation and the connection with the spatially distributed actuator system. The higher level is devoted to the modulation of the autowave propagation via another spatio temporal dynamics, producing suitable commands, under the form of the so-called Turing patterns, to control the lower level dynamics. This layer directly manages the sensor inputs and in this way a real time locomotion control strategy can be implemented. Of course, local feedback can be also present to emphasize the versatility of the approach.
The methodology introduced focuses on generating and controlling a number of different gait types in an hexapod insect-like robot. In the subsequent lectures it can be easily extended to other bio-inspired robots already built in our lab.

2 Locomotion as a spatio-temporal phenomenon

Neurobiologists agree on the fact that the neural architectures for locomotion control must have evolved in a hierarchical way. For this reason the common organisation of biological neural networks is commonly functionally studied as a hierarchical structure. In particular, the study is mostly performed by following some stereotypes both for invertebrates and for vertebrates. In general it is agreed that the Central Nervous System (CNS) must produce specific patterns of motor neuron impulses during a coordinated motion. As in Marsden et al. (1984) we can define a motor program as "a set of muscle commands which are structurated before a movement begins and which can be sent to the muscle with the correct timing so that the entire sequence is carried out in the absence of pheripheral feedback". The central hypothesis is that there is a neural pattern generator, within the CNS, that produces the basic motor programs (Wilson, 1972). Information derived from sensory inputs may modify the output of the pattern generator so as to adapt locomotion to the environment. More specifically, rhythmical movements that drive locomotion effectors (muscles), are triggered by a group of neurons, that can be called *Local Motion Generation Neurons* (LMGNs). They in turn are controlled by a higher level neural center, called *Centre of Command Neurons* (CNs), which fixes a particular locomotion scheme based either on specific signals coming from the CNs, or on feedback deriving from sensory inputs. In such a way the output of

the pattern generator can be modified so as to adapt locomotion to the environment (Stein, 1978; Calabrese, 1995).
Indeed many studies have been attempted to address specific pathways to try to unravel the organization and functions of the CPG in vertebrates and in invertebrates. Their conclusion is that the CPG's neural organization is more complex than needed to merely generate motor oscillations. In fact, for example, all the motor systems known possess several mechanisms to generate a given rhythm: this is clear in the sense that a given motor activity can be initiated directly by the CNS or by afferent signals from peripheral feedback. In fact all the CPGs are hybrid systems: they are also able to generate oscillations or plateau potentials over the firing threshold. This is a key issue for rhythm generation, but also for driving the transition among various types of locomotion (walking, running, swimming) (Pearson, 1993). In the sea snail *Aplysia*, for example, the interesting property of some command neurons is their ability to produce plateau potentials beyond the firing threshold, while in general rhythmical motor systems, neurons with intrinsic oscillatory dynamics, chemically of hormonally modulated, are commonly found.
Afferent feedback can modulate the intrinsic pre-programmed neural activity for a fine adaptive behaviour. In Figure 1 a schematic representation of the role of the CPG in the motor system is shown.

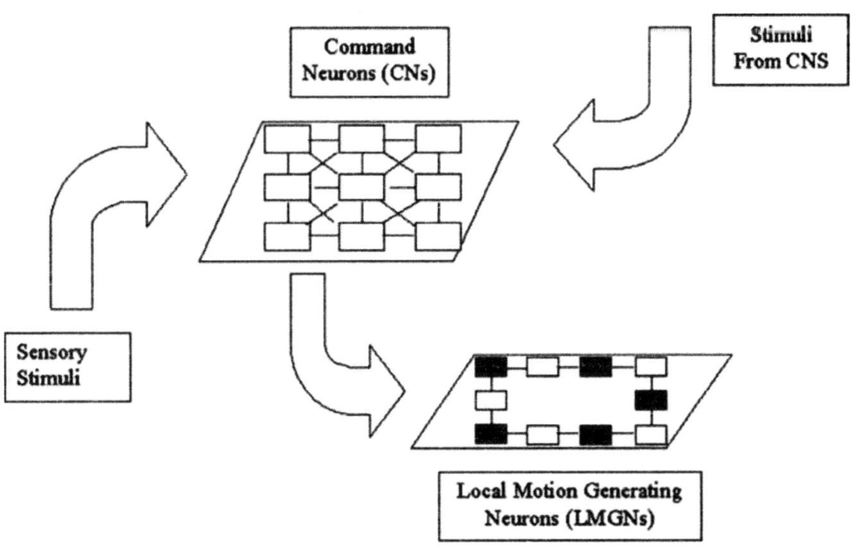

Figure 1. The Central Pattern Generator Scheme

From a behavioral point of view, the whole locomotion system (CPG) therefore appears to be a complex activator-inhibitor spatio-temporal system, characterized by a hierarchical organization in which a group of neurons (the CNs), due to sensory or central excitations, activate other groups of neurons (the LMGNs) that generate the appropriate timing signals for the type of locomotion induced by the CNs.
What previously stated by neurobiologists are working hypotheses, and, when applied to any specific motor system, may have to be modified to provide a more accurate theory concerning the function of that specific motor system.

We focus at deriving a simple but efficient model of the CPG. To this aim the following important features are to be cited: the dynamic of the single cell versus a population dynamic; the role of connections among the cells; the role of feedback in the system performance. These issues will be discussed in the following subsections.

2.1 Rhythmical activity as a result of self-organisation

Generally a specific motor program is performed by the concurrent activity of a population of neurons, and not by a single one. In neurobiology neural populations that are responsible for the motion pattern generation are often referred to as *neural oscillators*, since the organised dynamics useful to drive specific motions is nothing else than a triggered oscillation. Moreover the number of neurons that participate to the onset of a given neural rhythm is seldom known with a certain precision, since only a small fraction of neural cells is directly involved, with respect to the whole number of cells in a given neural site. On the other hand the complete model of a single neuron gives rise to a very complex dynamics, often showing chaotic motions. The high connectivity, typical of a neural tissue, gives rise to a very complex network of oscillators, showing a dynamics which happens to wander in hyperchaotic spatio-temporal motions characterised by low-power consumption, during the resting period, and however able to show highly organised periodic dynamics when excited by a particular afferent stimulus (Freeman, 1992). Since we want to model only organised structures, we do not need to implement complex models, but can restrict our interest to ensembles of low-order nonlinear oscillators whose main characteristics strictly resemble those ones shown by intrinsic neural signal processing. Therefore we, in this work, will refer to a model of neurons in the organised state, showing stable oscillations. However, the spatio-temporal wave-like dynamics met in neural processing possess original characteristics that make them heavily different from conventional spatio-temporal waves. These characteristics were observed during the huge number of experiments made in the early 50's by the Nobel prizes Hodgkin and Hunxley (Scott, 1995). They derived a quantitative description of the dynamics met in the isolated neural fibers of the giant squid and found out that the neural impulse can be considered as an "original" wave front, which has peculiar characteristics. First, a necessary condition for the onset of a neural impulse is to apply a suitable "over-threshold" potential, otherwise the neuron lies in a resting, non-spiking, state. Moreover, the time spent in the resting state is generally much greater than the firing time: therefore the neural dynamics can be modelled as a "slow-fast" dynamics. Once produced, the impulse propagates at constant speed along the axon, with constant amplitude and form during propagation. If two neural impulses collide in a neural tissue, they annihilate, rather than penetrate one another. On the contrary, if they meet together while propagating in the same direction, their wave fronts will fuse together and will synchronise. All these properties, finely included in the Hodgkin-Hunxely (H-H) dynamic neuron model, were proved subsequently in several measures both performed in laboratory and in living tissues (Scott, 1995). Moreover, the neural impulse does not save energy: since the neural tissue has all the properties of an active nonlinear medium, the impulse propagation takes place at the expenses of the energy locally released by the tissue which is used by the impulse itself to propagate towards the neighboring sites. These characteristics are commonly met in spatially distributed nonlinear dynamical subsystems coupled with diffusion laws. With these considerations, instead of looking for high order nonlinear equations to describe the dynamics of the neuron, in view of a circuit realisation, it was preferred to design nonlinear oscillators able to generate slow-fast dynamics, coupled one another by diffusive laws and showing the properties discussed above. Under these consideration, the design of a neural oscillator will mean for us the construction of a slow-fast nonlinear system which stands for an aggregate of neurons showing organised activity and obeying to the peculiar propagation characteristics mentioned before.

2.2 The role of the connections in the organization of neural dynamics

The single neuron activity is transferred to the other cells via suitable connections which realise neural population often in closed loop, ring-like configurations, as in the case of Nematodes (Niebur and Erdos, 1993). In the neural structures synaptic transmission can be realised in two main ways: chemical and electrical. The first one is perhaps the most common synaptic type and here the presynaptic termination transforms the electrical signal into a chemical one, and through a complex transformation, back into a post-synaptic potential. The second one takes place when two neurons establish links one another via high conducibility intercellular bridges (Shepherd). These connections are peculiar since they do not cause more delays in the signal propagation than the electrical signal propagation itself. For this reason such synapses are common in the neural circuits that are devoted to efficiently work at high speed. This is the case, for example, of the segmental ganglia of the crayfish that drive the escape reactions. In our model, we'll refer to this kind of synaptic links and will model electrical diffusion connections among the neurons.

2.3 The role of feedback in the system performance

There are several types of control loops in a CPG. The main are at a local and at a high level. In some cases motion is realised mostly in an open loop scheme, through commands sent directly from the Central Nervous System. At the high levels feedback has the role of selecting the suitable limb motion based on the environment information (Ghez et al., 1991). Therefore it has the complex role to organise all the single joints motions. This results in complex spatio-temporal dynamics to be imposed to the LMGNs in Figure 1, most often positioned next to the single joint actuators. In fact some other feedback loops operate directly connected to the terminal fibers most often to regulate the frequency, phase and amplitude of motoneural activity to adapt the particular movement to the irregularities of the environment. In this sense this type of afferent feedback is devoted to control the transition from one to the other phase of movement or to reinforce ongoing motor activity. This is the case of the cockroaches, where cuticular strain detectors are used to directly control, during the slow walking, the transition between the stance an the swing phase. In most of biological examples this task is performed by using sensors integrated into the terminal CPG elements, for example in the case of the swimming regulation of the lamprey (Grillner et al., 1991), or in the flight system in the locust (Pearson, 1993). In our work, the main emphasis is devoted to the part dedicated to the high control centres which generate the motion patterns, and we'll assume the information derived from sensory signals to act only at the high levels of the locomotion type planning. Therefore feedback effect at this stage is only occasional and devoted to plan a suitable motor program.

3 Modelling the CPG

The idea of using arrays of oscillators to build neural like activities to artificially generate locomotion patterns is not new. Since the early part of this century the idea of two half-centres coupled by reciprocal inhibition and able to show alternating activity was introduced, and during the 70s, ring-like structures containing two or more neurons were investigated for their emergent oscillatory dynamics. Also the gait in centipedes was modeled as driven by wave propagation. Moreover, even rings of oscillators, enriched with diagonal connections were seen as suitable models to produce gaits similar to those ones observed in tetrapods or hexapods. Unfortunately, most of such models did not find any physical realization, mainly since, at the beginning, physical devices were not able to realize the characteristics shown by these models. During the 70s, some mathematical models were introduced together with some software simulations. One decade later, the introduction of Artificial Neural Networks gave rise to a series of very interesting applications for the control learning of stepping machines, biologically inspired (Cruse et al., 1998). The interesting experiments reproduced had to stop to the software simulation.
The new strategy outlined in this lecture considers locomotion as a complex phenomenon, taking place in space and in time and involving a great number of variables. Such concepts can be applied referring to the analog CNN paradigm. Arrays of programmable electronic analog circuits can be easily realized, which can implement the CPG. Moreover one could not neglect the concept of an analog circuit realization with respect to digital implementation or even to software simulation. In fact the latter two solutions loose their efficiency as the number of variables and of actuators to be concurrently handled grows up. The analog implementation of autowaves and Turing patterns allows to realize locomotion as the solution of partial differential non linear equations in a spatial medium. Each joint of the robotic structure represents no longer an independent control variable, but the time course of a variable spatially dependent on the neighboring ones in a membrane-like structure.
Under this perspective attention has to be paid to look for mathematical structures that could serve as a model for continuous time, spatially distributed phenomena. One of the most famous and mostly used models in mathematical biology refers to the so-called Reaction-Diffusion equations, whose general form is the following:

$$\begin{aligned} \frac{\partial A}{\partial t} &= F_1(A,I) + D_A \nabla^2 A \\ \frac{\partial I}{\partial t} &= F_2(A,I) + D_I \nabla^2 I \end{aligned} \tag{1}$$

being: A and I two generalized variables standing for the chemical concentrations of the activator and inhibitor respectively, in a so-called activator-inhibitor mechanism suggested in Murray (1993), $F_1(A,I)$ and $F_2(A,I)$ non linear functions, D_A and D_I the diffusion coefficients,

$$\nabla^2 = \frac{\partial^2}{\partial x^2} + \frac{\partial^2}{\partial y^2} \tag{2}$$

the two-dimensional Laplacian operator. Such nonlinear PDEs are commonly used to model natural phenomena, among which self-sustained oscillations, met, for example, in bursting phenomena, or in morphogenetical pattern formation, and so on. In particular our interest will be focused on these two steady state spatio-temporal dynamics, i.e. Autowaves and Turing patterns.

3.1 Autowaves and Turing patterns

Wave-like self-sustained oscillating phenomena, possessing all the properties of autowaves (Krinsky, 1984), as well as steady-state patterns, can be obtained as solutions of eq.(1). The term *autowave* was firstly coined by R. V. Khorhlov, to indicate "autonomous waves". They represent a particular class of nonlinear waves, which propagate without forcing functions, in strongly nonlinear active mediums. Autowaves posses some typical characteristics, basically different from those of classical waves in conservative systems. Their shape remains constant during propagation, reflection and interference do not take place, while diffraction is a common property between classical waves and autowaves. A necessary condition for the onset of autowaves is: $D_A \cong D_I$. The key point is that all the characteristics of autowaves belong to the neural firing. Therefore the latter one can be artificially generated if a structure is designed so as to reproduce autowave fronts.
From a macroscopic point of view, in the simplest moving animals, among which, for example nematodes, but also in some mollusks like squids, some types of locomotion are directly induced by the propagation of an impulse

signal autowave-like, along the neuron axon. The soft body structure is able to synchronize with the traveling wave and a wave-like motion onsets. In more developed animals, like insects, from a high level point of view the autowave propagation can be still supposed to generate locomotion, but the neural structure has been improved with a highly organized and much more complex organization, where the locomotion types are pattern selected. In such systems the rhythmic movements are driven by the CPG: its capacity to generate also plateau potentials or oscillations is a key issue for gait generation, but also for driving the transition among various types of locomotion, such as walking, running or swimming (Pearson, 1993).
The generation of plateau potentials can be also modeled as solution of eq.(1) under the form of the so-called *Turing patterns* (Turing, 1952), usually met when chemicals react and diffuse in such a way as to produce steady-state heterogeneous spatial patterns of chemical concentration (Murray, 1993). In the case of pattern formation, the diffusion phenomenon takes place spatially: in particular, the activator is responsible of the initial instability into the medium, and the pattern formation starts; once this phase is completed, the inhibitor supplies stability. A necessary condition for such a phenomenon to take place is that: $D_A << D_I$.
As previously outlined, both autowaves and Turing patterns are solution of eq.(1). To design circuits able to reproduce such dynamics, particular attention was focused on CNNs Chua and Yang (1988); Chua and Roska (1993), since they are spatially distributed arrays of nonlinear analog computing cells, as outlined in the following Section. CNNs allow to efficiently implement biologically inspired motion. The CPG paradigm is realized by inducing autowaves and patterns in CNN arrays. Such spatio-temporal waveforms are exploited to drive suitable and self-organized motions in an ensemble of actuators that move some mechatronic devices. The most important aspect regards the coordination and control among these actuators, so as to realize a suitable control strategy. The approach to control non linear spatio-temporal phenomena, if conceived with traditional techniques, appears quite prohibitive: the CNN used in this application contains a lot of state variables, mutually coupled in such a way that some of them suitably drive the robot actuators. Once again the inspiration comes from the experimental neurobiological results. As well known, animals, for example insects, possess several types of locomotion, able to fit to nearly every type of environment and of work.
In this perspective, feedback plays a fundamental role. In fact, in animal's CPGs, feedback exists at least at two main levels: local and high level. Local level feedback uses local signals from sensors to control each single joint; the high level feedback takes into account a broader information from the en-

vironment to decide a particular or a suitable sequence of pre-programmed locomotion schemes for an efficient degree of adaptation. In this way, if some conditions are fulfilled, the higher control centres impose suitable reference signals on the local pattern generators so as to change locomotion type. Of course the strategy is implemented in real time, even in the simplest animals. Moreover, while the local control task is to slightly modulate the amplitude or to presently lock the leg swing or stance phase, the high level control has the more complex and delicate function to manage and organize a great number of actuators and to elaborate suitable control schemes which have also to result congruent with the dynamic of each single joint. For this reason the attention will be devoted in this article to the strategy used to implement the high control centres with the CNN approach.
While the locomotion generation is simply accomplished by using a CNN grid generating autowaves, the locomotion control is realized by using another equal CNN structure with different templates, able to generate Turing patterns. Each pattern in the steady state configuration, realizes a particular locomotion scheme, simply implemented by imposing particular topologies to the CNN pool generating autowaves. In this way the wave front propagation is easily and finely controlled in real time: this gives the opportunity to modulate all the local neural dynamics so as to finely control the transition among different locomotion types.
For example, trajectory tracking in a legged robot is realized by a "pattern flow" at the high control centres which translates into a modulation of the spatio-temporal dynamics of the autowaves in the CNN pool. To this condition corresponds a particular combination of the robot legs which move following the rhythm of the controlled wave fronts. The great advantage of the approach is that both patterns and autowaves are realized with the same CNN structure (Arena et al., 1998a). Therefore the same CNN analog device (Arena et al., 1998b) can implement the whole structure in real time and independently on the number of actuators involved in the locomotion. The details on the CNN basic architecture are reported in the following lecture.

Bibliography

P. Arena, M. Branciforte, and L. Fortuna. A CNN based experimental frame for patterns and autowaves. *Int. Jour. on Circuit Theory and Appls*, 26: 635–650, 1998a.

P. Arena, R. Caponetto, L. Fortuna, and L. Occhipinti. *Method and circuit for motion generation and control in an electromechanical multi-actuator system.* Europ. Patent No. 98830658.5, 1998b.

J. Ayers, J.H. Witting, and K. Safak. Development of a biomimetic underwater ambulatory robot: advantages of matching biomimetic control architecture with biomimetic actuators. In *Proc. of SPIE, Sensor fusion and decentralized control in robotic systems III*, volume 4190, Boston, Ma, Nov. 2000.

K. Berns. *The walking machine catalogue. http : //www.fzi.de/ipt/WMC/walking_machines_katalog.*

R. A. Brooks. *Cambrian Intelligence.* MIT Press, 1999.

R.L. Calabrese. Oscillation in motor pattern-generating networks. *Curr. Op. in Neurobiol.*, 5:816–823, 1995.

L. O. Chua and T. Roska. The CNN paradigm. *IEEE Transactions on Circuits and Systems*, 40:147–156, 1993.

L. O. Chua and L. Yang. Cellular Neural Networks: Theory. *IEEE Trans. on Circuits and Systems I*, 35:1257–1272, October 1988.

CLAWAR. *Prooceedings of the First International Symposium Climbing and Walking Robots.* Brussels, 26-28 November, 1998.

H. Cruse, T. Kindermann, M. Schumm, J. Dean, and J. Schmitz. Walknet a biologically inspired network to control six-legged walking. *Neural Networks*, 11:1435–1447, 1998.

J. Dean, T. Kindermann, J. Schmitz, M. Schumm, and H. Cruse. Control of walking in stick insect: from behavior and physiology to modeling. *Autonomous Robots*, 7:271–288, 1999.

F. Pfeiffer et al. The tum walking machine. In *Proc.* 5th *Int. Symp. on Robotics and Manufacturing*, 1994.

W. J. Freeman. Tutorial on neurobiology: from single neurons to brain chaos. *Int. Journ. of Bifurcation and Chaos*, 2(3):451–482, 1992.

C. Ghez, W. Hening, and J. Gordon. Oganization of voluntary movement. *Current Opinion in Neurobiology*, 1:664–671, 1991.

S. Grillner, P. Wallen, L. Brodin, and A. Lansner. Neuronal networks generating locomotor behavior in lamprey. *Ann. Rev. Neurosci.*, 14:169–200, 1991.

V.I. Krinsky. *Self-Organization: Autowaves and Structures Far from Equilibrium*, volume 9–18, chapter Autowaves: Results, Problems, Outlooks. Springer-verlag, berlin edition, 1984.

C.D. Marsden, J.C. Rothwell, and B.L. Day. The use of pheripheral feedback in the control of movement. *Trends in Neurosci*, 7:253–258, 1984.

J. D. Murray. *Mathematical biology.* Springer-Verlag, 1993.

E. Niebur and P. Erdos. Theory of the locomotion of nematodes: control of the somatic motor neurons by interneurons. *Math. Biosci.*, 118:51–82, 1993.

G.K. Pearson. Common principles of motor control in vertebrates and invertebrates. *Ann. Rev. Neurosci.*, 16:265–297, 1993.

A. Scott. *Stairway to the mind.* Springer-Verlag New York, 1995.

G. M. Shepherd. *Neurobiology.* Oxford Univ. Press, 1997.

P.S.G. Stein. Motor systems, with specific reference to the control of locomotion. *Ann. Rev. Neurosci*, pages 61–81, 1978.

A. M. Turing. The chemical basis of morphogenesis. *Phil. Trans. R. Soc. London*, B327:37–72, 1952.

D.M. Wilson. Genetic and sensory mechanisms for locomotion and orientation in animals. *Am. Sci.*, 60:358–365, 1972.

Design of CPGs via spatial distributed non linear dynamical systems

Paolo Arena

Dipartimento di Ingegneria Elettrica Elettronica e dei Sistemi,
Universitá degli Studi di Catania, Catania, Italy

1 Cellular Neural Networks basics

The classical CNN architecture, in the particular case where each cell is defined as a nonlinear first order circuit is shown in Figure 1, in which u_{ij}, y_{ij} and x_{ij} are the input, the output and the state variable of the cell C_{ij} respectively; the cell non linearity lies in the relation between the state and the output variables by the Piece Wise Linear (PWL) equation (see Figure 1(c)):

$$y_{ij} = f(x_{ij}) = 0.5 \cdot (| x_{ij} + 1 | - | x_{ij} - 1 |)$$

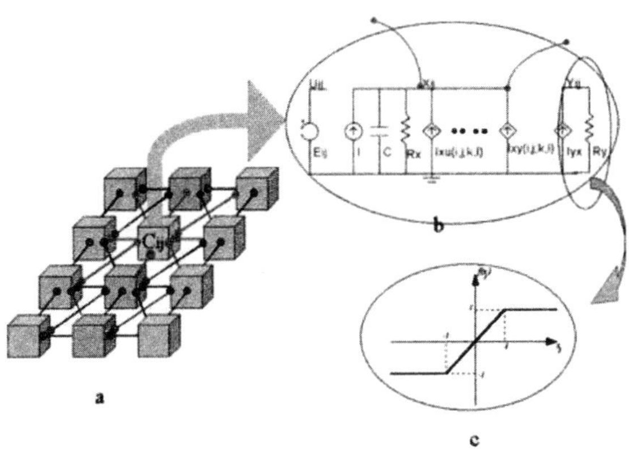

Figure 1. The CNN architecture: (a): the overall structure, showing local connections among the cells; (b):the basic cell structure, where red lines indicate the neighboring cell influences; (c): the CNN cell nonlinearity

The CNN architecture is classically defined as a two-dimensional array of MxN identical cells arranged in a rectangular grid, as depicted in Figure 1(a). Each cell (see Figure 1(b)) mutually interacts with its nearest neighbors by means of the voltage controlled current sources $I_{xy}(i,j;k,l) = A(i,j;k,l)y_{kl}$ and $I_{xu}(i,j;k,l) = B(i,j;k,l)u_{kl}$. The coefficients $A(i,j;k,l)$ and $B(i,j;k,l)$ are known as the *cloning templates:* if they are equal for each cell, they are called *space-invariant* templates and take on constant values. The CNN is described by the state equations of all cells:

$$C\dot{x}_{ij} = -\frac{1}{R_x}x_{ij}(t) + \sum_{C(r,s)\in N_\sigma(i,j)} A(i,j;r,s)y_{rs} + \sum_{C(r,s)\in N_\sigma(i,j)} B(i,j;r,s)u_{rs} + I \qquad (1)$$

with $1 \leq i \leq M, \quad 1 \leq j \leq N$
where

$$N_\sigma(i,j) = \{C(r,s) \mid max(\mid r-i \mid, \mid s-j \mid) \leq \sigma\}$$

with

$$1 \leq r \leq M, \quad 1 \leq s \leq N$$

is the $\sigma -$ *neighborhood* and

$$x_{ij}(0) = x_{ij0}; \quad x_{ij0} \leq 1; \quad C > 0; R_x > 0.$$

The classical CNN structure can be easily generalized in many ways, leading to the most complex CNN architecture: the so-called CNN Universal Machine (CNNUM) (Roska and Chua, 1993). Basically, it consists of an electronic architecture in which the analog CNN has been completed by digital logic sections. Here the term dual computing has been introduced. In this architecture the templates play the role of the instructions of a CPU, i.e. the templates determine the task that the CNNUM processor must accomplish. So, in the CNNUM, the programmability (i.e. the ability to change the templates in order to execute the various steps of a dual algorithm) is a central issue. The easy VLSI implementation (Chua and Yang, 1988; Chua et al., 1991) is due to some key features of CNNs with respect to traditional artificial neural systems. One of these, of course, is the local connectivity, while another is the fact that the cells are mainly identical. This advantage has permitted the development of many CNN real implementations (Manganaro et al., 1999). From the previous considerations, the CNN paradigm is well suited to describe locally interconnected, simple

dynamical systems showing a lattice-like structure. On the other hand, the emulation of PDE solutions requires the consideration of the evolution time of each variable, its position in the lattice and its interactions deriving from the space-distributed structure of the whole system.
Indeed, the numerical solution of PDEs always requires a spatial discretization, leading to the transformation of a PDE into a number of ODEs. Therefore the original space-continuous system is mapped into an array of elementary, discrete interacting systems, making the CNN paradigm a natural tool to emulate in real-time spatio-temporal phenomena, such as those ones described by the solutions of PDEs. This led to define a particular CNN architecture for emulating the Reaction-Diffusion equations: the so-called Reaction-Diffusion CNN (RD-CNN) (Chua, 1995).
The spatial discretization and the template definition are the two main steps to "electronically model" a PDE. Of course, it can be possible to start from the CNN, i.e. to design a CNN able to generate spatio-temporal signals that behaviorally represent solutions typically shown by nonlinear PDEs. In such a way, once the suitable template set has been derived, the analytical solution of some particular, space-discretized PDEs can be approximated by the CNN state equations.

2 RD-CNNs for autowave and pattern generation

Several papers appeared in literature on how to use CNNs to generate both autowaves and Turing patterns (Chua, 1995; Goras et al., 1995). The conditions for the onset of these phenomena can be translated in sufficient conditions for the parameters of the CNN cell, as well as of the whole CNN array. In this Section a RD-CNN array of second order nonlinear cells is adopted, where both autowaves and Turing patterns are obtained. The model is introduced through some definitions typically used in CNN terminology, and some propositions are reported, whose proof is reported in the referenced literature.

2.1 The CNN Model

Proposition 1: The basic cell dynamics of an RD-CNN for generating both autowaves and Turing patterns is a second-order system defined as in

Arena et al. (1998a):

$$\begin{aligned} \dot{x}_{1;i,j} &= -x_{1;i,j} + (1 + \mu + \epsilon) y_{1;i,j} - s y_{2;i,j} + i_1 + \\ &+ D_1 \cdot (y_{1;i+1,j} + y_{1;i-1,j} + y_{1;i,j-1} + y_{1;i,j+1} - 4 y_{1;i,j}); \end{aligned} \quad (2)$$

$$\begin{aligned} \dot{x}_{2;i,j} &= -x_{2;i,j} + s y_{1;i,j} + (1 + \mu - \epsilon) y_{2;i,j} + i_2 \\ &+ D_2 \cdot (y_{2;i+1,j} + y_{2;i-1,j} + y_{2;i,j-1} + y_{2;i,j+1} - 4 y_{2;i,j}); \\ & 1 \leq i \leq M; \quad 1 \leq j \leq N; \end{aligned} \quad (3)$$

with

$$\begin{aligned} y_{l;i,j} = 0.5 \cdot (| x_{l;i,j} + 1 | - | x_{l;i,j} - 1 |) \\ l = 1, 2; \quad 1 \leq i \leq M; \quad 1 \leq j \leq N. \end{aligned} \quad (4)$$

The terms inside parentheses, at the right hand side of equations (2), represent the discretised version of the two dimensional laplacian operator (eq.2 of lecture 1).

Proposition 2: Let us consider a MxN CNN array with a cell defined as in Proposition 1. If it is assumed for each cell:

$$\mu = 0.7, \ \epsilon = 0, \ s_1 = s_2 = s = 1, \ i_1 = -0.3, \ i_2 = 0.3$$

sufficient conditions are satisfied such that each isolated cell shows a slow-fast limit cycle, while the corresponding RD-CNN generates autowaves. If:

$$\mu = -0.6, \ \epsilon = 1.82, \ s_1 = 2, \ s_2 = 2.5, \ i_1 = i_2 = 0,$$

the cell satisfies all the conditions for pattern formation (Arena et al., 1998a, 1997).
The new RD-CNN can be now defined, using the formalism introduced in Chua and Yang (1988).

Definition 1: The state model of the new two-layer CNN with constant templates is defined by the following equations:

$$\dot{x}_{ij} = -x_{ij} + A * y_{ij} + B * u_{ij} + I$$

where $x_{ij} = [x_{1;i,j} x_{2;i,j}]'$, $y_{ij} = [y_{1;i,j} y_{2;i,j}]'$ and $u_{ij} = [u_{1;i,j} u_{2;i,j}]'$ are the state, the output and the input of the CNN respectively while A, B and I are the feedback, control and bias templates respectively and the operator

'$*$' is recalled in the following definition.
Definition 2(Convolution Operator):For any template T it holds:

$$T * v_{ij} = \sum_{C(k,l)\in N_\sigma(i,j)} T(k-i, l-j)v_{kl}.$$

Proposition 3: Taking into account the previous propositions and definitions, the cloning templates for the RD-CNN are characterized as follows (Arena et al., 1997):

$$A = \begin{pmatrix} A_{11} & A_{12} \\ A_{21} & A_{22} \end{pmatrix}; \quad B = 0; \quad I = \begin{pmatrix} i_1 \\ i_2 \end{pmatrix}; \tag{5}$$

where:

$$A_{11} = \begin{pmatrix} 0 & D_1 & 0 \\ D_1 & -4D_1 + \mu + \epsilon + 1 & D_1 \\ 0 & D_1 & 0 \end{pmatrix}; \tag{6}$$

$$A_{22} = \begin{pmatrix} 0 & D_2 & 0 \\ D_2 & -4D_2 + \mu - \epsilon + 1 & D_2 \\ 0 & D_2 & 0 \end{pmatrix};$$

$$A_{12} = \begin{pmatrix} 0 & 0 & 0 \\ 0 & -s_1 & 0 \\ 0 & 0 & 0 \end{pmatrix}; \quad A_{21} = \begin{pmatrix} 0 & 0 & 0 \\ 0 & s_2 & 0 \\ 0 & 0 & 0 \end{pmatrix}.$$

In particular, for autowave formation, the following conditions hold:

$$D_1 = D_2 = 0.1,$$

while for pattern formation it is assumed:

$$D_1 = 0.01, \ D_2 = 1.$$

In both cases the so-called *Zero-Flux (Neumann)* boundary conditions (Chua, 1995) have been assumed.

The robustness of this CNN model to parametric uncertainties and noise is also guaranteed (Arena et al., 1998a), in view of its hardware implementation.

2.2 Remarks

The importance of this approach is that the same structure can implement the autowave propagation or the pattern formation only by suitably modulating its parameters: the cloning template coefficients and the overall structure remaining unaffected. Since both of the phenomena, autowaves and Turing patterns, are autonomous phenomena, the B template is zero. Therefore the whole phenomenon is characterized by the A and I template values. As it can be derived from Proposition 1 and Proposition 2, the design of the CNN basic cell led to a two-layer CNN. This means that each cell is a second order nonlinear circuit. This implies that the A and I templates are non longer simple matrices, but block matrices, that, in order to match the CNN general form (Def. 1), have to be ordered as in eq.(5). Each sub-block matrix coefficient derives by ordering the parameters of the cell model (eq.(2)). For example the middle element of matrix A_{11} represents the influence on the state variable $x_{1;i,j}$ of its own output $y_{1;i,j}$. Therefore the parameters belonging to this middle element are: $1+\mu+\epsilon-4D_1$ (eq.(6)). Moreover, since the connection between each cell in the first layer and the neighboring cell in the same layer is only diffusive and only in the compass directions, the other A_{11} coefficients are D_1 or zero. The same considerations hold for the template A_{22}, for the second layer of the CNN. The A_{12} and A_{21} template values define the interaction between the two layers of each CNN cell. Since there is no interaction among a cell of a layer and the neighbors of the other layer, only the middle parameters are non zero, and in particular they are the s parameters from eq.(2). Another important issue to outline is that while eq.(1) seen in the first lecture is a nonlinear PDE, the CNN paradigm refers to nonlinear lattices, that have been widely demonstrated to be the space discretized version of the corresponding nonlinear PDEs (Scott, 1999). In this way it can be stated that the solutions obtained by the RD-CNNs represent the discrete space approximation of the corresponding RD-PDEs seen in eq.(1) of the first lecture.

3 Rexabot: Reaction-Diffusion hexapod walking robot

In this section the RD-CNN approach for the real time generation and control of an insect-like autonomous walking robot is described. Rexabot,shown in Figure 2, was entirely realized in our laboratory. The body was built by using different sizes of aluminum rods. The whole weight of the mechanical part is about one Kilogram. The robot dimensions are $25x15x15$ cm.

Each leg has two degrees of freedom and is moved by two servomotors:

Figure 2. Rexabot during its locomotion

one drives the vertical position of the leg foot, in such a way as to realize the stance and lift phases, while the other one drives the rotation of the leg to realize the locomotion swing phase. The servomotors used in this prototype are low cost model airplane position controllable motors, able to provide a maximum torque of 3.5 Kg/cm. They were chosen to make the whole structure able to carry one further Kilogram, enough for the batteries, control and driving circuits, so as to make Rexabot completely autonomous.
It is to be outlined that Rexabot structure is quite similar to many other legged robots. In fact the focus here is not on the mechanical structure or actuator system, but on the new gait generation and control methodology. In particular the leg motion is modeled as a steady state oscillation: when the leg is in the stance phase, i.e. when it is holding the body weight, the motion speed is lower than during the swing phase, in which the leg is detached from the ground. In such a way the motion can be represented by a slow-fast limit cycle, like that one designed previously (see Proposition 2). Moreover, the robot kinematics has been designed so that each leg dynamics can be driven by the two state variables of one CNN cell.
The locomotion pattern generation is related to the CPG discussed previously. Namely two main neural circuits will be explored: that one generating the signals to drive the motion actuators (the LMGNs)and the CNs, devoted to generate suitable stimuli to control the LMGNs in the locomotion generation task. Referring to Figure 1 seen in the first lecture, autowaves will be used for the locomotion generation. In particular a CNN ring generating autowaves implements the LMGNs if its state variables directly drive the leg motors. The LMGN dynamics is controlled by the CNs via suitable pattern

configurations. The LMGN layer is constituted by a ring of 12 cells. In our case, if particular cells in the CNN ring are suitably connected to the legs, as the wave fronts propagate through the ring, motion takes place in each leg.
If the legs are synchronized one another, thanks to a suitable connection cells-legs controlled by the CNs, it is possible to implement different types of gaits. Therefore a relevant role is assumed by the CNs which provide a "locomotion scheme" through the generation of a specific pattern for the LMGNs and therefore for the actuator system.
The CNs are modeled by a RD-CNN, generating Turing patterns, whose outputs are plateau signals, therefore steady state commands, which reflect the scheme for the right connection between the cells generating autowaves and the actuator system robot legs in order to realize a particular locomotion type. Moreover the CNs must possess a discrete quantity of steady state configurations, in order to realize a number of different locomotion types. Finally, the particular steady state condition, corresponding to a particular locomotion type, has to be directly controlled by different types of sensory inputs, and possibly in parallel. With such considerations, and recalling the basic properties of Turing patterns, their application to model the CNs dynamics appears very natural for the following characteristics of Turing patterns that can be easily obtained with CNNs:

- in a CNN composed also of a little number of cells, it is possible to obtain a number of different steady state patterns. For example, in Arena et al. (1998b) it has been analytically proven that in a RD-CNN 3x3, eight different stable pattern configurations are available;
- it is easy to drive the network towards the desired pattern by setting the suitable boundary and initial conditions;
- a steady state configuration reached can be changed to reach another desired one by forcing the state values of some cells.

The structure depicted configures as a distributed neural network able to reach a finite number of steady state patterns as a function of some control inputs to the network cells. Therefore it appears to be a natural way to model the CNs.

4 implementation of the control strategy

The classical fast gait of insects (for example the tripod walking of cockroaches) is shown in Figure 3(b).

Here white bars represent the swing phase, i.e. the the phase when leg is detached from ground, or the stance phase, when the legs are on ground (black part) (Pearson, 1976). Referring to Figure 1 of the first

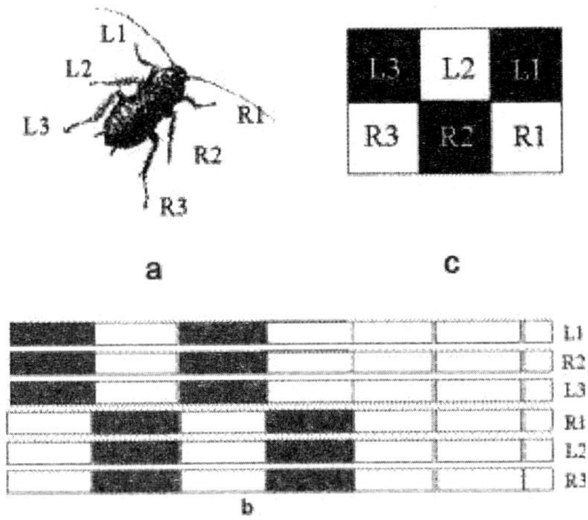

Figure 3. (a): Labels for each leg of the hexapod; (b) a scheme for the fast gait; (c) the Turing pattern realizing the fast gait.

lecture, as said before, the LMGNs are mapped on the CNN, continuously generating autowave fronts, while the CNs implement a particular Turing pattern configuration.

As depicted in Figure 3, a RD-CNN 2x3 checkerboard Turing pattern was used to modulate the LMGNs for the generation of the specific locomotion: "fast gait". The following strategy was used: in order to implement the CNs by using RD-CNN generating Turing patterns, each cell in the two rows of the 2x3 RD-CNN matrix directly corresponds to a leg. Therefore such cells have been labeled with the corresponding leg name. Now it is imposed that the same autowave front drives the legs corresponding to the cells with the same color in the Turing pattern configuration. In fact the steady state signals A_{11}, A_{22} and A_{13} assume low voltages, representing a synchronous motion of the associated legs $L3$, $R2$ and $L1$, respectively. On the contrary, the steady state signals A_{12}, A_{21} and A_{23} assume high voltage, representing a synchronous motion of the associated legs $R3$, $L2$ and $R1$, respectively. Therefore the legs: $L3$, $R2$ and $L1$ will move obeying to the same wave front, while the remaining ones are connected to a cell which results to be opposite in phase with respect to the cell used to drive the first leg tripode combination. In such a way this type of locomotion can be easily mapped into a RD-CNN.

According to the schematic representation of the insect legs of Figure 3, the

particular locomotion "fast gait" is accomplished by connecting the insect tripode $L1 - R2 - L3$ to the same RD-CNN cell, for example A_0 showing autowaves, and the second tripode $R1 - L2 - R3$ to a CNN cell which shows an autowave with a phase shift of 180 degrees with respect to A_0.
Indeed, the type of gait realized by insects is not only the "fast gait" depicted up to now, but also the "medium" and the "slow" gait, represented in Figure 4.

Figure 4. Schemes for the medium gait (a) and slow gait (b)

Our aim is to establish, a suitable strategy to connest the LMGNs to the robot legs to realize also these other types of locomotion. It is evident that in biology the spiking frequency of neurons cannot vary as a mere consequence of the speed of walking. Therefore an hypothesis is that the LMGNs can activate different pathways, which involve a different number of neurons. Let us make the following working hypotheses, considering the scheme of Figure 5.

- There are a number of neurons connected in a loop, in which, for simplicity, each neuron is modulated only by the preceding one and is able to affect the spiking of the following one. This can be easily realized in our CNN array: in fact, a suitable way to have a unique autowave front continuously propagating into the neuron loop is to connect the last cell of the array to the first one.
- Let us think that one given neuron A_0 is connected to the muscle system of a leg so as to generate motion once for each firing. Let us also imagine that one neuron fires: its spike "travels" along the neural loop exciting subsequently all the neurons in the loop.
- Under these conditions let us think to modify the number of neurons within the loop, without changing the structure of each cell. Since

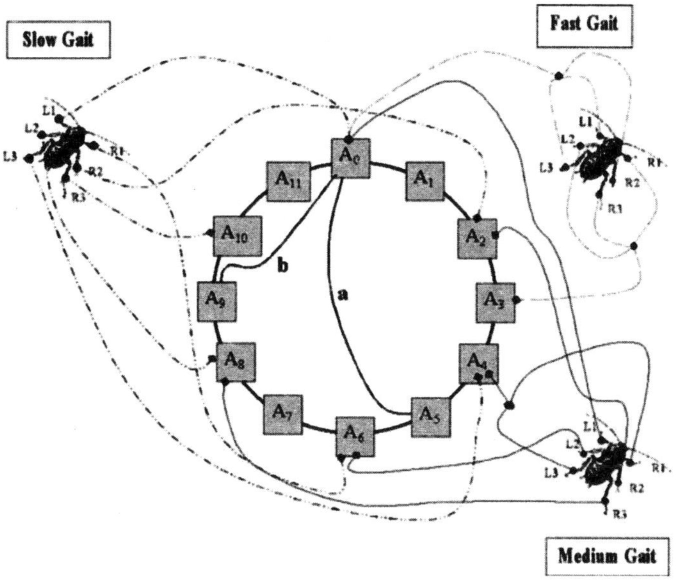

Figure 5. Connections between LMGN and the legs of the hexapod to realize the fast, the medium and the slow gait

> the cell parameters are unchanged, their oscillation speed does not vary, if each cell is isolated from the neighboring ones. The diffusive connection among the cells acts so as to organize among each other all the single oscillations. In fact each neuron fires only if it receives suitable stimulation from the previous cell, subsequently passing the stimulation, under the form of a wave front, to the following cell in the ring. The main result of this "emerging property" is that, the fewer the number of neurons within the ring, the faster the speed of propagation of the wave along the ring. Of course, the maximum frequency of the wave propagation is equal to the frequency of each isolated cell. It was experimentally proven that a minimum number of neurons must be present into the ring in order to see a wave front propagation; this is clear if we think that the spatial wave length must be smaller than the ring dimension.

As a result, in order to want to implement the gait by using a chain of neurons oscillating at a given frequency, the fast gait would require a smaller number of neurons in the chain with respect to the medium gait and, of course, to the slow one. The strategy already outlined is depicted in Fig-

ure 5, which shows the CNN ring and the connections cell-leg. The fast gait is realized by connecting the tripode $L1 - R2 - L3$ to the cell A_0 and the other tripode to the cell A_3 which has a phase shift of 180 degrees. Furthermore, for symmetry reasons, two other cells must be used and A_5 has to be connected to A_0 to close the neuron chain. The loop used for the autowave propagation is, for the fast gait, $A_0 - A_5$ and the path denoted with "a". The medium gait derives from the fast gait if a certain phase delay takes place among the legs belonging to the same tripode and if the legs $L3$ and $R1$ move synchronously as pointed out in Figure 4(a). Once again we can refer to Figure 5 and see the connections among the loop $A_0 - A_7$–b and the robot legs. In fact now more cells are needed to make up the ring: in these conditions the autowave front will take more time to visit twice the same cell. Therefore the frequency of activation of each leg is lower than one in the case of the fast gait. The conclusion is that there is a phase delay among the cells that depends on the number of neurons involved in the ring. It is to be again outlined that this characteristic is peculiar of this implementation of a reaction-diffusion system, since we do not modify the parameters of the setup; each oscillator in the ring, due to the diffusion phenomenon, does not fire until it does not receive suitable input from the neighbors. Therefore the firing frequency in the ring depends on its neuron number. Such adaptation is the key issue to easily obtain different behaviours on the same structure.
Similar considerations hold for the slow gait (see Figure 4(b)). In this case the whole loop $A_0 - A_{11}$ is used and the dot-dot-solid lines depict the connections between the LMGNs and the robot legs. In this simple way all the three gait locomotion types can be efficiently implemented by using the CNN frame. As an example, in Figure6 a series of snapshots, depicting the Rexabot while performing one "fast gait" cycle is shown.

4.1 Feedback Control of the Locomotion Direction

Another interesting feature that can be easily realized with the frame proposed regards the feedback control of the direction of the robot during its locomotion. Let us consider once again the scheme of Figure 5. If the robot is walking "fast", analogously to the biological case, a change of direction, for example in the right hand side, is realized making one of the legs of the tripode $R1 - L2 - R3$, namely $R1$, to make a longer motion with respect to the others. In this case, the leg $R1$, instead of being moved synchronously with the other two legs by the cell A_3, is connected also to the cell A_2. In this way R_1 is allowed to move first than the others, and in conclusion it makes a wider angle and realizes the direction change. Of course, the

Figure 6. Snapshots depicting, in the sequence indicated by the letters, one "fast gait" cycle for the hexapod prototype; (a-d): motion of the tripod L_1, R_2, L_3; (e-h): motion of the other one

desired steering angle can be obtained with one or more subsequent gaits. Another steering strategy, which is indeed the one implemented in our case, is to freeze the swinging motion of the middle leg, say the right one, to realize the right turn.

Up to now the attention has been focused on the realization of LMGNs and their connection to the leg system so as to realize different types of gaits. The role of the CNs was restricted to the generation of a specific Turing pattern. However its importance becomes fully evident considering the implementation of other types of motion in hexapods, like swimming; these can be represented like patterns as well. In this case Turing patterns, implemented in the CNs, assume a relevant role in deciding the particular locomotion type under feedback of from CNS or from environment as indicated in Figure1 shown in the first lecture.

In this paper a 2x3 CNN has been implemented. It was proved in Arena et al. (1997) that such a structure is able to generate up to three stable patterns, which can be used to implement two other locomotion types. Of course, insects posses several different gait configurations, and CNNs are able to implement all of them under different steady state pattern configurations in the CNs. Once again, local connectivity and modularity allow to enlarge the array dimension when necessary.

The CNN approach to generate the CNs dynamics is appealing because the transition from a given pattern to another steady state configuration can be easily obtained by forcing the state of a small number of cells. For example, the initial conditions for the checkerboard need the imposition of only one state variable to guide the whole structure (12 state variables). The

same considerations hold in larger structures (for example in a 3x3 CNN, which has 18 state variables). This is another example that presents this phenomenon as a consequence of synergistic effects, in which a variable, the so-called *regulative variable* is able to make the whole complex system to reach a given order (Haken, 1981). These considerations acquire more relevance since one given stable configuration can be obtained by choosing one or more regulative variables.
The result is that, the more the number of regulative variables used concurrently, the faster the achievement of the steady state configuration. This is in perfect agreement with the strategy employed, in which each regulative variable can be connected to a different sensor.
For example, another classical type of locomotion type is "swim" Here the legs $R1$, $R2$ and $R3$ are moved synchronously and in opposition of phase with respect to the legs of the other side of the animal body. This type of pattern can be easily obtained with the frame proposed, as shown in Figure 7, in which the dashed cells indicate initial conditions which are around zero.

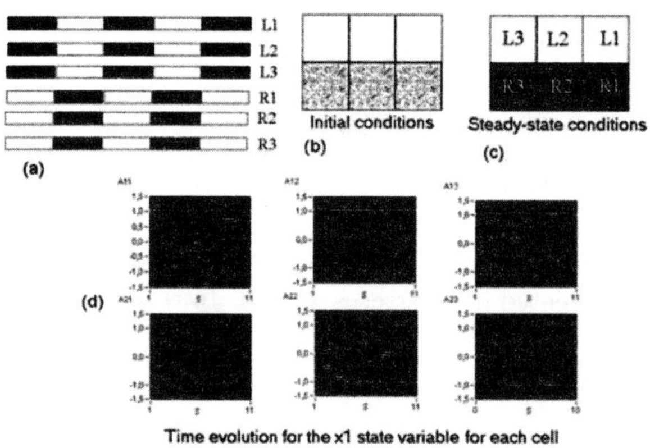

Figure 7. Formation of a strip pattern which corresponds to the "swim" locomotion; (a): scheme of the "swim" pattern; (b): initial condition for the CNN array; (c): steady state conditions; (d): transient for the state variable x_1 for the CNN cells

Just to emphasize the strength of the approach proposed, we can gain the transition between the locomotion "swim" to the locomotion "fast gait", which is accomplished for instance when an insect leaves water to reach the dry land. The transition between these two types of locomotion is depicted in Figure 8, where the sensory input, which controls the transition between the two types of locomotion, directly affects only the state variable x_1 of the A_{22} cell; this variable in this case assumes the role of regulative variable.

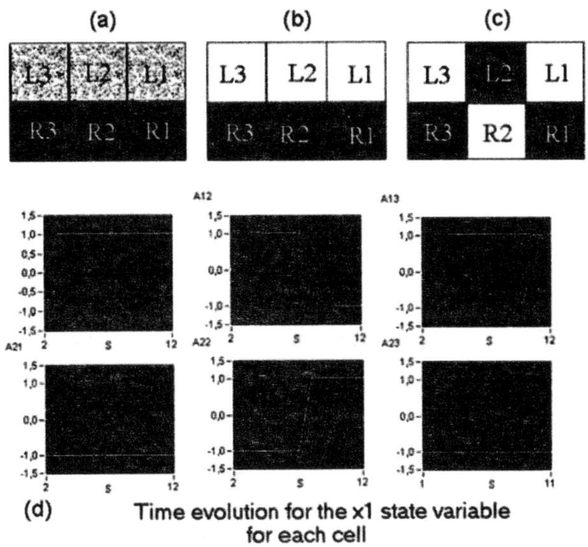

Figure 8. Pattern formation and further evolution after perturbing the reached steady state conditions of the A_{22} cell. This correspond to the migration from the swim to the fast gait motion. (a) Initial conditions, (b) steady state conditions; (c) new steady state conditions after perturbation; (d) time evolution for the state variable x_1 of the CNN cells.

Therefore, the transition from the swimming configuration of Figure 8b, to the fast gait condition of Figure 8c, is realized following an analog signal flow from only one sensor to the network of CNs.

4.2 Experiments on Fault Tolerance: an Example of Emerging Property

The intrinsic robustness of the spatio-temporal phenomena treated in this paper gives often rise to unexpected emerging properties. Let us suppose that the fast gait walking is being reproduced: the robot is able to walk in the alternating tripode type. One might think that the setup built could be very sensitive to the circuit faults: i.e. the splitting or cutting of the ring, or the damaging of one cell could completely destroy the wave

propagation. This is not exactly true. Indeed the locomotion rhythm is lost in the sense that, for example in the fast gait, the time between the end of each cycle and the beginning of the following one can vary. Due to its diversity, each cell disconnected will oscillate with its own period. The ring connection, realized via a diffusion template, synchronizes all the oscillators to a unique rhythm. When the ring is split such rhythm is more or less affected. Another thing could be observed when splitting the ring: different spurious wave fronts can be found. In this case different pseudo-movements can be appreciated, and complex locomotion patterns emerge, i.e. the robot generally succeeds in the locomotion, but the "order" of propagation seen in the intact ring is lost. This behaviour is very similar to experiments on some insects giving rise to some affections in the intact locomotion and to the loss of the phase symmetry among the moving legs.
The previous phenomena are presented in the sequence of pictures in Figure 9, where the cutting of the ring does not extinguish locomotion, but produces a degradation of the intact locomotion cycle. This represents a clear example of "emerging property" in self organizing dynamics in spatial distributed systems showing complex behaviour.

Bibliography

P. Arena, R. Caponetto, L. Fortuna, and G. Manganaro. Cellular neural networks to explore complexity. *Soft Computing Research Journal*, 1(3): 120–136, 1997.

P. Arena, S. Baglio, L. Fortuna, and G. Manganaro. Self organization in a two-layer CNN. *IEEE Trans. on Circuits and Systems - Part I*, 45(2): 157–162, 1998a.

P. Arena, M. Branciforte, and L. Fortuna. A CNN based experimental frame for patterns and autowaves. *Int. Jour. on Circuit Theory and Appls*, 26: 635–650, 1998b.

L. O. Chua and L. Yang. Cellular Neural Networks: Theory. *IEEE Trans. on Circuits and Systems I*, 35:1257–1272, October 1988.

L.O. Chua. Special issue on nonlinear waves, patterns and spatio-temporal chaos. *IEEE Trans. on Circuits and Systems - Part I*, 42 (10), 1995.

L.O. Chua, L. Yang, and K.R. Krieg. Signal processing using cellular neural networks. *Journal of VLSI Signal processing*, 3:25–51, 1991.

L. Goras, L. O. Chua, and D.M.W. Leenaerts. Turing Patterns in CNNs-part I: Once over lightly. *IEEE Trans. on Circuits and Systems - Part I*, 42:602–611, 1995.

H. Haken. *Erfolgsgeheimnisse der Natur.* Deutsche Verlags-Stoccarda, 1981.

G. Manganaro, P. Arena, and L. Fortuna. *Cellular Neural Networks: Chaos, Complexity and VLSI processing.* Springer-Verlag, 1999.

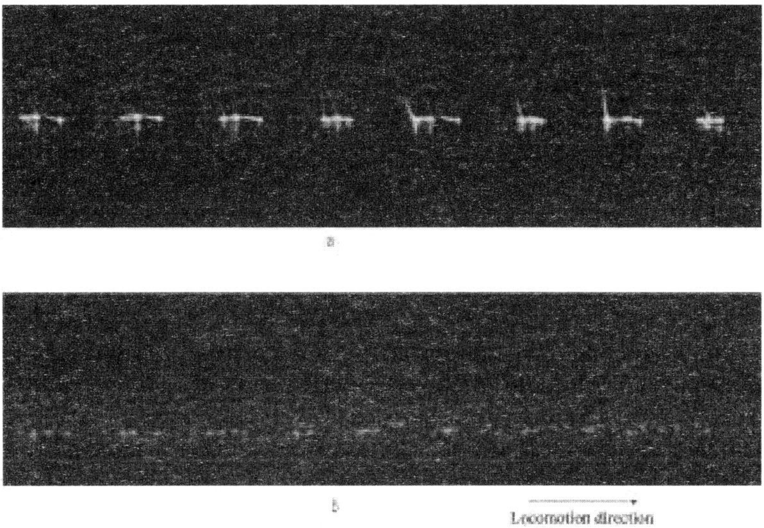

Figure 9. (a) regular locomotion leg cycles and (b) irregular leg cycles due to the CNN ring damaging. The image sequences were obtained by recording, by a long exposition photo, the light emitted by some LEDs applied to each foot of the hexapode. The red color outlines the front leg trajectory, the green the mid leg, the yellow the hind leg

G. K. Pearson. The control of locomotion. *Sci. Am.*, 235(6):72–86, 1976.

T. Roska and L. O. Chua. The CNN universal machine: an analogic array computer. *Trans.on Circuits and Systems - Part II*, 40:163–173, 1993.

A. Scott. *Nonlinear Science.* Oxford Univ. Press, 1999.

Realization of bio-inspired locomotion machines via nonlinear dynamical circuits

Paolo Arena

Dipartimento di Ingegneria Elettrica Elettronica e dei Sistemi,
Universitá degli Studi di Catania, Catania, Italy

Abstract In the previous lecture some design guidelines for CPGs by means of CNNs were given, giving particular attention to the realization of the gaits in a hexapod structure. In this lecture it will be shonw that the same spatial-temporal dynamics can be also used to obtain patterns for other kinds of bio-inspired moving machines. For the sake of clarity, in the following section the basic cell dynamical model, formally identical to that one used in the previous lecture, is briefly recalled.

1 The Two-layer Reaction-Diffusion CNN

In this section a two-layer linear CNN is recalled, able to show self-organization phenomena, useful for motion control purposes. It can be directly defined by introducing the structure of the single cell together with the numerical values of templates.

$$\begin{aligned} \dot{x}_{1;i,j} &= -x_{1;i,j} + (1+\mu)y_{1;i,j} - sy_{2;i,j} + i_1 + \\ D_1(y_{1;i+1,j} &+ y_{1;i-1,j} + y_{1;i,j-1} + y_{1;i,j+1} - 4y_{1;i,j}); \\ \dot{x}_{2;i,j} &= -x_{2;i,j} + sy_{1;i,j} + (1+\mu)y_{2;i,j} + i_2 + \\ D_2(y_{2;i+1,j} &+ y_{2;i-1,j} + y_{2;i,j-1} + y_{2;i,j+1} - 4y_{2;i,j}); \end{aligned} \tag{1}$$

with

$$y_i = 0.5 \cdot (\mid x_i + 1 \mid - \mid x_i - 1 \mid) \tag{2}$$

and

$$and \quad i = 0, 1, \cdots, M-1, \quad j = 0, 1, \cdots, N-1$$

Some propositions were proved in Arena et al. (1998a) and experimental results were presented in Arena et al. (1999) to show that the above system, while used as a cell in a MxN CNN array is able to show autonomous wave propagation for the following parameter set: $\mu = 0.7$, $\epsilon = 0$, $s_1 = s_2 = 1$,

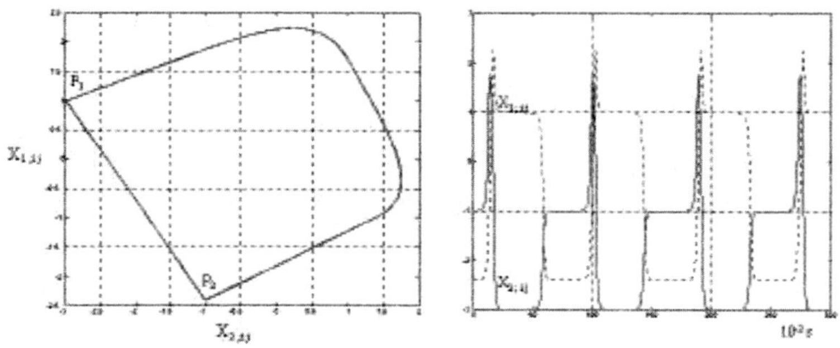

Figure 1. Slow-fast limit cycle and time trends for the two state variables of the single CNN cell.

$i_1 = -0.3$ and $i_2 = 0.3$, $D_1 = D_2 = 0.1$. The terms D_i represent the diffusion coefficient of the discrete Laplacian template to modulate local interactions among cells. The relations just written represent the equations of a two-layer autonomous CNN (Chua and Yang, 1988). From these parameter values it is straightforward to derive the templates values that "program" the whole CNN, already reported in Arena et al. (1997).

The simulation of this CNN structure, with the so-called *Zero- Flux (Neumann)* boundary conditions (Chua et al., 1995), gave rise to several complex phenomena of wave propagation, fully explained in Arena et al. (1998a). Moreover, in Arena et al. (1998b) a discrete device analog frame has been introduced and some measures on complex phenomena have been fully described. The slow-fast dynamics of the state variables of the single CNN cell is reported in Figure 1 together with the slow-fast limit cycle described by the two state variables of each neuron. In particular the two points P_1 and P_2 represent the slow dynamics points, the line and the curve in between the fast dynamics. As outlined before, each oscillating cell, locally connected to the other ones via diffusion couplings, is able to show autowave propagation, depicted in Figure 2 for a CNN 5x5.

In the following part of the lecture the computational model of CNNs generating autowaves will be applied to model the structures of some CPGs well known in neurobiology, together with interesting applications to some mechatronic devices suitably built. In all the cases considered in the following, for the considerations made in the previous Sections, it will be assumed that the basic motions for biological locomotion generation will be caused by autowaves. Therefore the structure and the topology of the CNN used to enable autowave propagation will be varied according to the needs of the

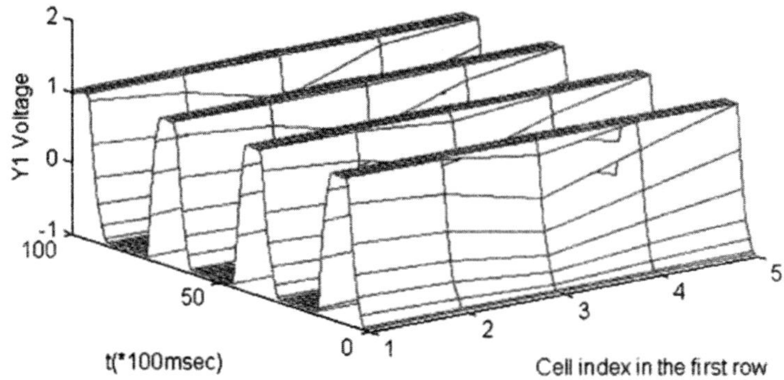

Figure 2. Autowave propagating through the CNN matrix rows.

particular realisation of the CPG under consideration.

2 the CPG in nematodes

Simple nervous systems in primitive animals are very useful to understand the organisation and the main functional aspects of biological neural systems. Electrophysiological experiments, linked with anatomical knowledge show that in some species of nematodes, such as in *Ascaris lumbricoides*, the motor neurons transmit electrotonical signals producing a wave of neural excitation; this is directly translated in a muscular wave, travelling at the same speed as the neural one. More recently, a lot of experimental work was performed on another nematode, the so-called *Caenorhabditis Elegans*, which helped to solve its neural map. This nematode has exactly 302 neurons, connected by about 5000 chemical and 2000 electrical synapses. In Niebur and Erdos (1993) and references therein, a complete model of the neural circuit is derived together with some computer simulations carried on linear models based on PDEs. It is shown that in the C. Elegans the motor nervous system is able to globally control the somatic muscular system by using only electrotonic spreads of potential, i.e. signals propagate through neurons with electrical potentials which can take on continuous values.
As regards the scheme of functional neural circuits, the main conclusions are that the C. Elegans has two main neural circuits, one for backward motion and one for forward motion. The two circuits have practically the same structure and make use of the same triggering neurons which directly

innervate the dorsal and the ventral muscle system. The two neural systems are constituted of 4 neural groups, each one possessing up to about 23 neurons, able to show organised electrotonic activity. They receive input stimuli directly from the two main neural groups: the *Nerve Ring* and the *Caudal Centre*. The former is the most complicated neural structure in the nematode. It is a densely connected region of toroidal shape, while the latter, simpler one, is situated in the posterior part of the body and connected to the Nerve Ring with a number of nervous bundles which run along the body axis. To these bundles are connected the somatomuscolar motor neurons in the ventral cord. These last ones directly drive the undulatory motion of the somatic muscles. From these considerations the nerve ring can be assumed to give the suitable locomotion patterns depending on afferent stimuli, while the caudal centre is responsible for the local motion generation.

2.1 The Worm-bot

This subsection is devoted to describe a biologically inspired robotic structure, where both the neurobiological issues and the hardware setup described previously will be used to realise artificial locomotion. In Figure 3 a picture of the robot is reported. It consists of five rings (R_1 to R_5), which simulate the rings of the nematode. Each ring is built-up of two legs which are actuated by a dc servomotor. The control signal is a square wave whose period ranges from 1.2 to 2.0 ms. To each period in between corresponds a particular angle of the motor shaft ranging between 0° and 180°.

Figure 3. The worm-bot.

The robot structure is very simple: each leg has only one degree of freedom which is the elevation angle from the ground. Referring to Figure 3, the feedforward locomotion is realised when the legs rotate clockwise around their axis; the opposite rotation does not cause locomotion. In fact each leg is made of two parts joint together with a knee-like hinge. During the clockwise rotation (ring R_1 and R_4 in Figure 3), the so-called stance phase, the two parts of the leg form a unique rigid support to realise locomotion, while during the opposite rotation (R_3 in Figure 3), the swing phase, the two parts can rotate one another and no motion of the robot body is realised. The knee-like motion is realised using a simple return spring, outlined in Figure 4. The five rings thus built are linked together via a plastic sheet in such a way as to resemble the body of a nematode.

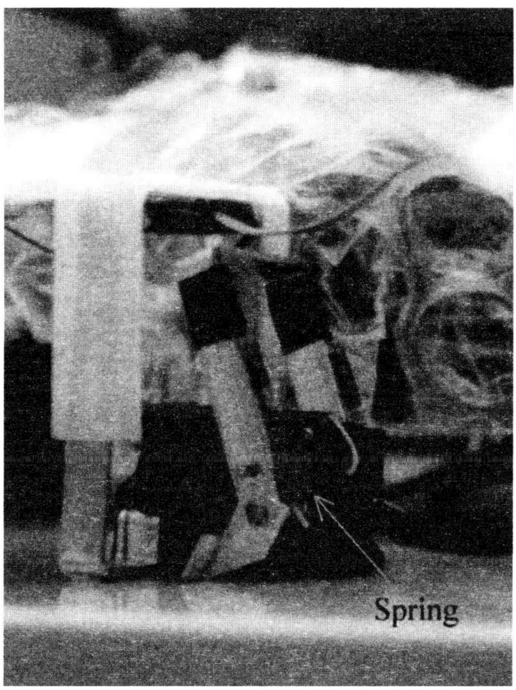

Figure 4. The worm-bot leg detail.

As regards the use of the CNN cells to realise the robot motion, the basic idea was to translate the slow-fast limit cycle into a cycle "stance-swing" of the robot legs. To this aim the control signal to each servomotor is one of the two outputs of a neural oscillator previously described. In particular the points P_1 and P_2 in Figure 1 represent the extreme positions for the stance phase, while the curve $P_2 \rightarrow P_1$ the swing one. The structure of the network able to realise to overall organization among the various rings of the robot is reported in the next Subsection.

2.2 Realization of the nematode CPG by CNNs

As previously outlined, the main neural circuit to be considered is the Nerve ring, which controls the whole pattern generation for undulatory locomotion. The other circuits are mainly devoted to the local coordination. In particular, let us suppose that a given stimulus has already been established in order to program a particular type of locomotion, for example the forward motion. In this case, an autowave front has to be generated, which is able to make the whole structure move forwards. To this aim a CNN ring was built. The scheme can be seen in Figure 5.

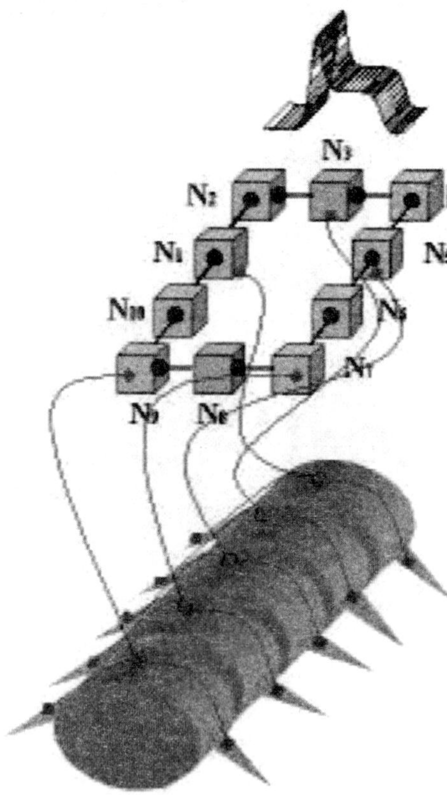

Figure 5. Scheme of the CPG for the locomotion generation in the wormbot.

Suitable initial conditions are set (Arena et al., 1998a) in such a way that an autowave front will propagate in a particular direction. In this

case, once onset, it will subsequently visit all the cells in the ring. So, the unique autowave front will excite subsequently all the servomotors which will describe a complete stance-swing oscillation. Therefore all the neurons in the ring will stay in their resting phase, (slow part of the dynamics), and only the neuron visited by the wave front will "fire" (fast dynamics). In such a way, if each ring in the robot is assigned to a particular neuron in the CNN ring, the wave propagation will drive consecutively all the rings of the robot, realising the locomotion. In particular, if the autowave propagation corresponds to a direction going from the back to the front side of the robot, the feedforward locomotion is realised. By using this strategy the speed variation can simply be realised. Referring to the scheme in Figure 5, we have twice the number of neurons in the ring with respect to the number of rings in the robot. Moreover, it is unsuitable to vary the frequency of oscillation for each neuron, because, referring to its hardware implementation, it would mean to vary the component values. A simple way to make the structure to go faster is to allow the neurons to be connected not subsequently, but one bridging another. The scheme in Figure 5 represents the slow locomotion, while, if direct connections $N_1 - N_3 - N_5 - N_7 - N_9$ are allowed, a "fast" undulatory locomotion takes place.
The CNN implementation of the CPG allows also some simple ways for the realisation of a trajectory control for the worm-bot. In fact, as outlined in the previous sections, the CNN structure used to generate the autowave fronts is autonomous. Therefore additional exogenous inputs to some given cells can be efficiently used to represent inhomogeneities into the medium of propagation and therefore to determine the propagation failure in the spatial region corresponding to the selected cells. In this simple mechanical structure the couple of legs corresponding to each ring is driven by the output of a cell. If in each ring two servo motors are placed, each one driving the position of one leg of the ring, and the autowave is allowed to propagate in a matrix having two rows, the CNN inputs can be used to make the autowave to propagate along the two legs of each single ring of the wormbot. The CNN inputs, suitably placed, prevent the autowave from propagating in some cells that in such a way are frozen. The corresponding legs are prevented from moving, while the opposite ones move regularly, realizing the direction change. The strength of the approach lies in the fact that autowaves do not suffer from reflection phenomena, caused from boundaries as well as from inhomogeneities in the medium (represented by the input signals). Therefore the whole wave propagation is not affected by reflection waves when inputs are used to modulate the robot trajectory.

The strategy is depicted in Figure 6 and Figure 7. The former figure shows the input mask to the CNN. The black color stand for a "high" input, while white stands for "low" input. This pattern causes the autowave propagation to follow the trajectory depicted in Figure 7. It can be seen that in correspondence to the "high" input levels the autowave does not propagate, and therefore no motion takes place in the corresponding legs.

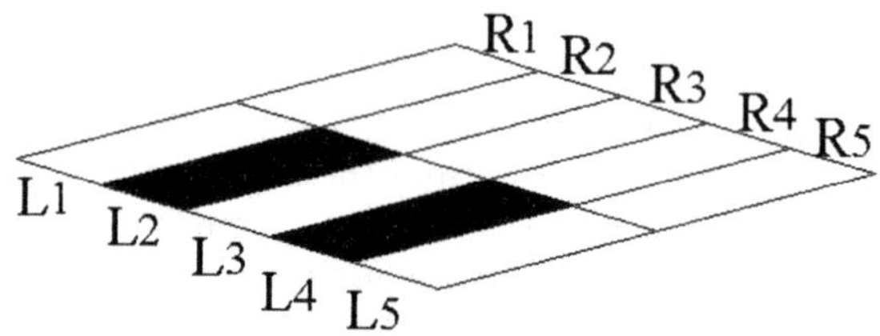

Figure 6. Input pattern to the CNN 2x5 to control the locomotion direction.

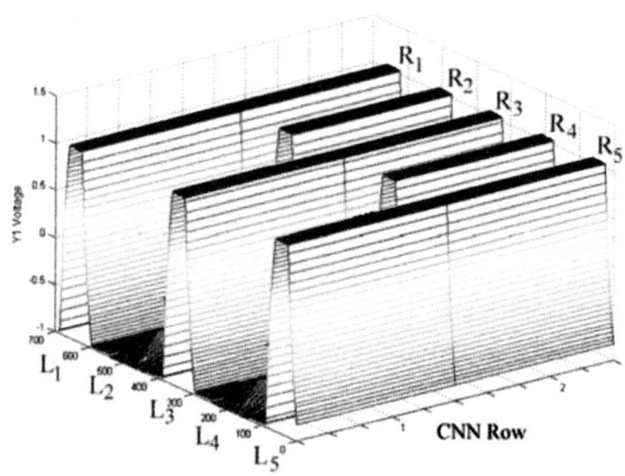

Figure 7. Autowave trajectory modification due to the input pattern of Figure 6.

Of course, the trajectory can be easily programmed by imposing a suitable image as a "mask" at the CNN input layer: this mask can be modified in real time to make the structure to adapt its path to the particular environment. The flexibility of the approach and the possibility to control the locomotion direction also in this simple example, make the approach particularly appealing. Moreover the whole strategy can be implemented in analog VLSI technique, which allows also the possibility to be suitably programmed at the input layer. In this way, the whole architecture assumes the features of a powerful soft-computing architecture for real time trajectory control.

3 Swimming generation in the Lamprey

The lamprey has been suggested as a good animal model for the study of CPGs, since it has far fewer cells that mammals, and some aspects of their circuitry can be elucidated electro physiologically. Much knowledge has been gained about circuits that produce locomotion in the lamprey nervous system. The lamprey spinal cord has about 100 segments, and each segment has about 1200 neurons. During locomotion four main cells types appear to be active: approximately 50 cross-caudal interneurons, 5 inhibitory interneurons and 50 excitatory interneurons per hemi-segment, plus 50 lateral interneurons in the entire spinal cord (Selverston, 1993). The lamprey nervous system can be roughly divided into a Cerebral Trunk (CT) and a Spinal Cord (SC), built-up of a certain number of Spinal Cord Segments (SCS) (Grillner et al., 1991). Like fishes, lamprey swims by progressively contracting its muscles via undulatory motions from head to tail. This is generally done by proper impulses coming from the CT circuits, but it has been discovered that also each SCS can initiate a movement which produces a coordinated activity with neighbored segments. Other studies confirmed that the CT can initiate swimming, but leaves the local coordination among the SCSs, and the muscles to local neural groups, which have the capability to react also to local sensorial stimuli. The CT is connected via long axons to the segments in order to drive the activation of one side of the body and the concurrent inhibition of the other side of each SCS to coordinate the oscillatory motion. One key factor is that the some particular SCSs have long axons to other particular SCSs along the SC. They of course posses short connections to adjacent SCS to diffuse local motion. These are roughly the main characteristics of the lamprey neurobiology (Grillner, 1996). At this point, what the neurobiologists draw about the functional aspects of the intricate neural system, is that in general the SCS next to the CT receive the initial stimuli. Therefore the segments next to the lamprey head tend to move faster than the other SCSs. These ones, once received the stimuli, tend to initiate their motions, but with a phase delay which, propagating along the body, gives rise to an homogeneous progressive wave. As regards the swimming speed, the time interval of the activation of neighbored SCSs is constant for a given wave, but its period can go from about 3 to 0.1 seconds for the fastest jumps. These main guidelines inspired our work on how to generate swimming in an underwater lamprey-like robot. In fact the swimming system as a whole represents a complex reaction-diffusion system, which has similar characteristics in a lot of biological cases, among which the examples presented before. Starting from the point that we consider the neuron dynamics as "always active", we can model it as a nonlinear oscillator locally connected to other equals cells by diffusion rules. As a consequence the model can be once again described by using a CNN array to generate suitable signals to drive swimming in a lamprey robot prototype.

In particular the following neurobiological issues are fundamental for our model:

- the lamprey's body contains a single wavelength of oscillation at any given time;
- some particular SCSs have long axons to other particular SCS along the SC;
- each segment is able to generate bursting activity in response to certain sensory inputs.

For the first issue we can model the undulatory nervous system as an array of nonlinear oscillators, interacting mainly via local connections.

In Figure 8 a realisation of this scheme is presented. Here the SC of the Lamprey is represented as a series of SCSs locally connected.

Figure 8. A schematic representation of the lamprey swimming nervous system

The last neuron in the caudal SCS is directly connected to the first one in the first SCS. Therefore we have realised a ring configuration. It can be experimentally shown that in the circuit realisation of the CNN ring if some spurious wave fronts are generated, they will annihilate and only one autowave will propagate along the ring, realising in this way the first issue. Each SCS is made up of a number of locally connected neurons. Therefore the whole scheme can be represented as in Figure 9.

If suitable initial conditions are imposed from the CT, an autowave front will onset: this front will "visit" consecutively all the neurons in each SCS. In particular we are supposing that all the neurons in each SCS innervate the muscles in the SCS, thus locomotion can be realised by exciting all or a fraction of the neurons in each SCS. In this condition an undulatory motion along the lamprey body is realised. Figure 10 shows the same configuration,

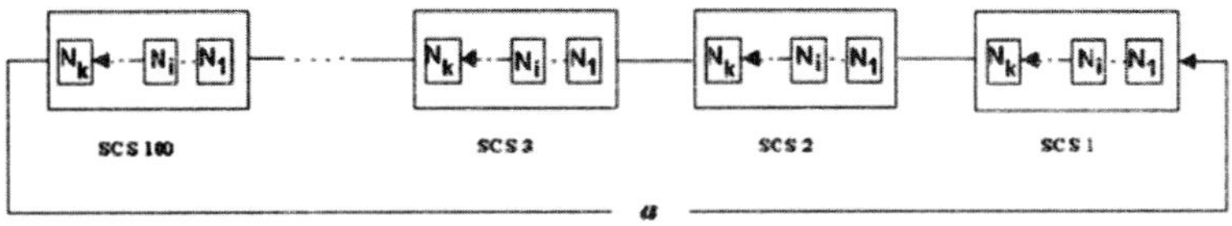

Figure 9. A scheme for the realisation of the slow swimming scheme.

but now there are some particular neurons that propagate their signals also to the first neuron in the following and previous SCS.

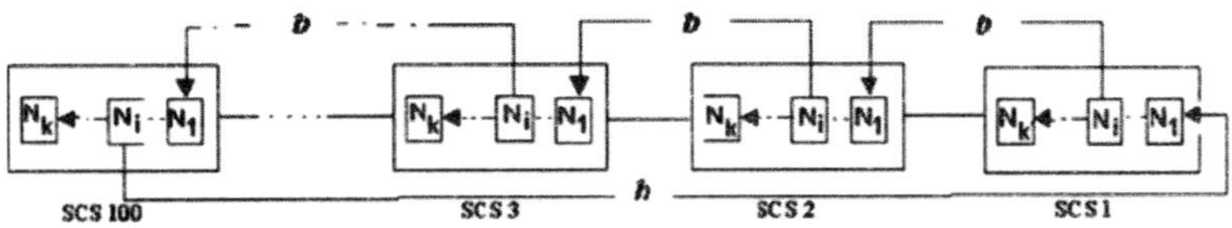

Figure 10. A scheme for the realisation of the medium swimming scheme.

In this condition the autowave front will visit a fraction of all the neurons in the SC; therefore the period spent by the wave front to come back to the first SCS will be smaller. Of course the fastest speed is realised in Figure 11, where only one neuron per SCS is able to fire. In such a way the speed is increased simply by decreasing the neuron number involved in to the progressing firing.

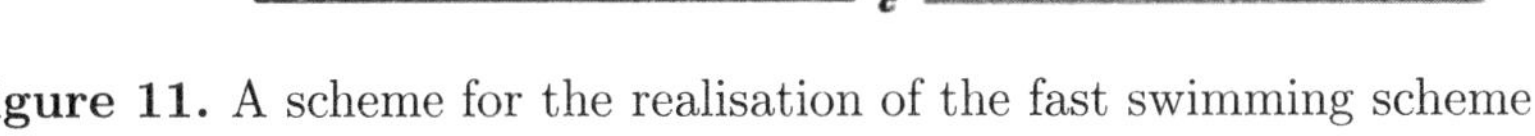

Figure 11. A scheme for the realisation of the fast swimming scheme

Of course in these figures the CT has not been reported. Indeed its main function is:

- to initiate motions, i.e. to impose the suitable initial conditions;
- to provide suitable stimuli to inhibit or excite proper connections to realise swimming speed variation;
- to manage the sensory inputs and organise the overall locomotion.

Figure 12. The Lamprey robot structure.

At this point also the second neurobiological issue has been used. The third one comes out directly from our particular implementation of autowaves in CNNs, that allows each cell to be locally excited, for example by a sensor placed in a particular segment. In this case an autowave front will initiate from that segment, in full accordance with neurobiological findings, and will propagate to the whole body. In this unifying approach, the CNN dynamics is exactly the same as that one used to implement locomotion by autowaves in the previous experiments. Therefore the dynamics of the two state variables for each CNN cell used in this example are reported again in Figure 1. They indeed realise the local reactive slow-fast dynamics, while, diffusing to the neighbouring cells realise autowave propagation. The CNN ring used for the lamprey swimming coordination was made-up of 12 cells, while our lamprey prototype consists of 4 segments plus the caudal and the front ones, not actuated. We used a set of three neurons for each segment, in such a way as to give the possibility to implement the fast, medium and slow way of swimming.

3.1 The Lamprey Robot Mechanical Structure

The Lamprey robot, as already outlined, is made up of four actuated segments plus the head and the tail. The material chosen for the realisation is aluminium, thanks to its qualities of easy manufacturing and a relative degree sturdiness, enough for our purposes. The robot structure is reported in Figure 12.

It roughly consists of an aluminium backbone, made up of four vertebrae having elliptical shape of axes 16cm and 12cm long and 2 mm thick, and rods (20 cm long and 1 cm diameter each) inserted in the centre of each vertebra to realise the whole structure.

Motion in the segment is realised by means of pneumatic valves which drive some pneumatic muscles. In particular each segment is able to perform horizontal motion, since two muscles work as a flexor-extensor couple. Each muscle is controlled directly by a pneumatic valve, whose driving signal comes from the state variable of a particular CNN cell, showing oscillatory dynamics, coupled by diffusion templates in the ring already discussed. Each vertebra, which accomodates the valves, is also able to rotate, in order to allow spiral motions that realise downward and upward swimming, as in can be derived from the particular of the robot structure shown in Figure 13.

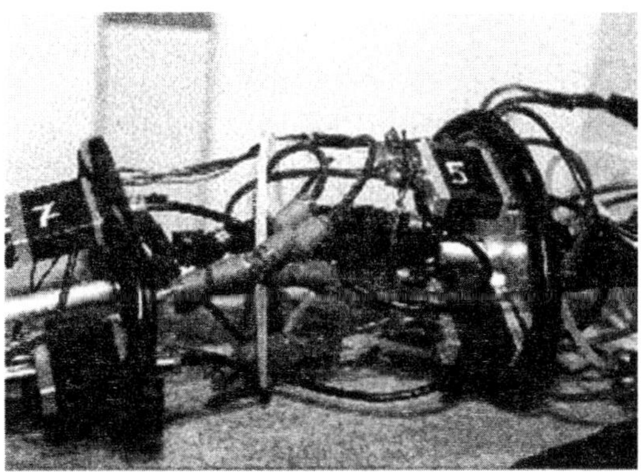

Figure 13. A segment of the robot structure.

The rotation of each vertebra with respect to the rod is realised via a ball bearing. The rods are connected each other via some joints that allow axial movements. The head of the lamprey robot is built up of plastic pipes ,suitably arranged to be waterproof. Inside the head the CNN cells for the swimming pattern generation are placed together with some other circuitry that drive the actuators according to the CNN cells state variables. At present the lamprey tail is realised with aluminium even if in the future it will be realised with plastic material so that produce a smoother swimming. Finally the robot was covered with waterproof elastic material typically used by skin-divers. The sheet was modelled, glued and stitched to allow axial and spiral movements due to its natural elasticity, but to prevent slipping

Table 1. Main characteristics of the pneumatic valves used.

Producer: Matrix S.p.a	Filtering degree: Max 50 micron
Power supply : 24 VDC 6 10%	Max Frequency: 220 Hz
Single Input Single Output	Weight: 25g
Pressure range: 2-8 bar	Life: /500 M/cycles
Tractable fluids: shallow air , neutral gas $(-10° + 50°C)$	Code: MX821100C2-24.

along the vertebrae. Finally, between the mechanical structure and the artificial skin, a thin plastic film was put to further isolate the structure from unavoidable perspiration.

3.2 The Swimming Actuation

The actuation of swimming was realised by using a pneumatic approach, consisting of Air Muscles driven by electrical valves. In particular we have used the McKibben pneumatic artificial air muscles. They consist of an internal rubber tube supported by braided cords that are attached at both ends to realise a tendon-like structure. When the air goes inside the rubber tube, the high pressure pushes against the external shell, and tends to increase its volume. According to non extensibility (or very high longitudinal stiffness) of the threads in the braided mesh shell, the actuator shortens according to its volume increase and produces tension to the applied mechanical load. This physical configuration causes McKibben muscles to have a variable-stiffness spring-like characteristics, nonlinear passive elasticity, physical flexibility, and very light weight compared to other kinds of artificial actuators. To realise our prototype 20 mm muscles were used; they are able to contract, under a pressure of 6 bars and a load of 5 Kg, of a percentage of the 34% of their length. Each muscle is driven by a digital pneumatic valve whose main characteristics are reported in Table 1.

In Figure 14 a complete portrait of the lamprey while swimming a in pool is shown.

Bibliography

P. Arena, R. Caponetto, L. Fortuna, and G. Manganaro. Cellular neural networks to explore complexity. *Soft Computing Research Journal*, 1(3): 120–136, 1997.

P. Arena, S. Baglio, L. Fortuna, and G. Manganaro. Self organization in a

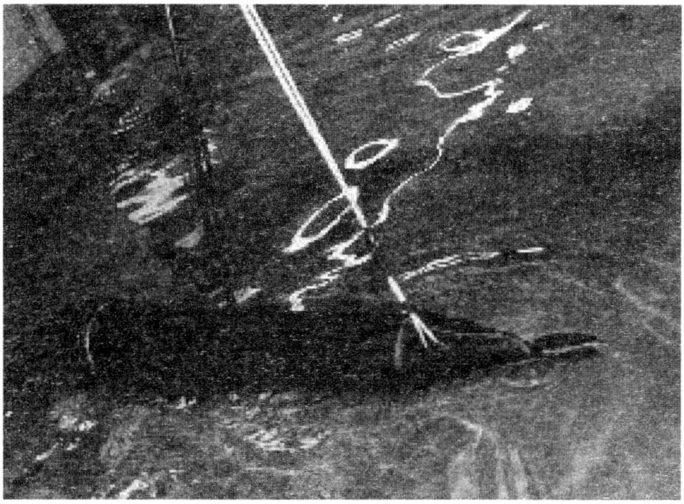

Figure 14. The Lamprey robot while swimming in a pool.

two-layer CNN. *IEEE Trans. on Circuits and Systems - Part I*, 45(2): 157–162, 1998a.

P. Arena, M. Branciforte, and L. Fortuna. A CNN based experimental frame for patterns and autowaves. *Int. Jour. on Circuit Theory and Appls*, 26: 635–650, 1998b.

P. Arena, L. Fortuna, and M. Branciforte. Reaction-diffusion CNN algorithms to generate and control artificial locomotion. *IEEE Transactions on Circuits and Systems I*, 46(2):253–260, 1999.

L. O. Chua and L. Yang. Cellular Neural Networks: Theory. *IEEE Trans. on Circuits and Systems I*, 35:1257–1272, October 1988.

L.O. Chua, M. Hasler, G.S. Moschytz, and J. Neirynck. Autonomous cellular neural networks: A unified paradigm for pattern formation and active wave propagation. *IEEE Trans. on Circuits and Systems I*, 42:559–577, 1995.

S. Grillner. Neural networks in vertebrate locomotion. *Sci. Am*, pages 64–69, Jan 1996.

S. Grillner, P. Wallen, L. Brodin, and A. Lansner. Neuronal networks generating locomotor behavior in lamprey. *Ann. Rev. Neurosci.*, 14:169–200, 1991.

E. Niebur and P. Erdos. Theory of the locomotion of nematodes: control of the somatic motor neurons by interneurons. *Math. Biosci.*, 118:51–82, 1993.

A. I. Selverston. Modeling of neural circuits: What have we learned? *ann. Rev. Neurosci*, 16:531–546, 1993.

Using robots to model biological behaviour

Barbara Webb

Institute of Perception, Action and Behaviour
School of Informatics, University of Edinburgh

Abstract Robots can be used to instantiate and test hypotheses about biological systems. This approach to modelling can be described by a number of dimensions: relevance to biology; the level of representation; generality of the mechanisms; the amount of abstraction; the accuracy of the model; how well it matches the behaviour; and what medium is used to construct the model. This helps to clarify the potential advantages of this methodology for understanding how behaviour emerges from interactions between the animal, its task and the environment.

1 Introduction

What does it mean to say a robot is used as a model? Consider the process of explanation and prediction described in figure 1.

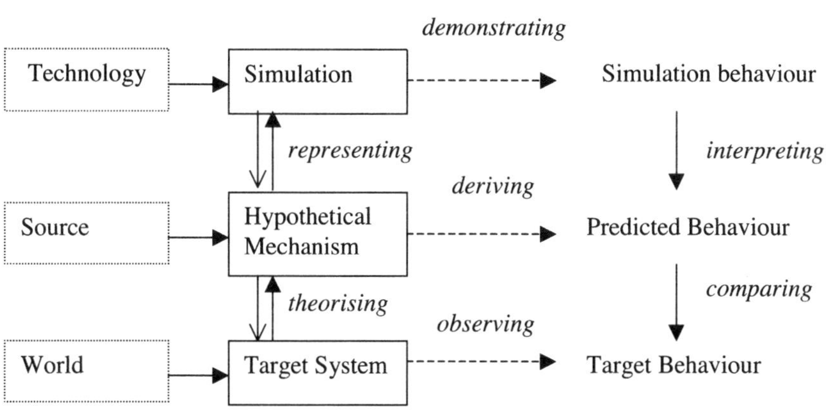

Figure 1. The process of explanation

To better explain each part of this diagram, take the example that will be discussed at greater length in the following chapters. Our target - selected from many possible examples the world - is the sound localisation behaviour of the cricket. Our hypothesis is that rather simple neural circuitry might be sufficient to enable the female cricket to respond to the pattern and direction of the male cricket calling song. One source for this idea is the simple internal wiring diagrams that result in interesting behaviour in Braitenberg's 'vehicle' thought-experiments Braitenberg (1984). To see what our hypothesis predicts we build a simulation of the neural processes and run it using robot technology, to see if it produces behaviour like the cricket. Insofar as it fails to match we might conclude our hypothesis is not yet sufficient to explain cricket behaviour. Chan and Tidwell (1993) usefully summarise this process as: we theorise that a system is of type T, and construct an analogous system to T, to see if it behaves like the target system.

The term 'model' is, confusingly, often used for different parts of figure 1. The target can be considered a 'model', as it is meant to be representative of a class of systems in the world; in biology, a particular animal is often described as a 'model species' for investigating a general problem (e.g. the fruitfly for genetics). The hypothesis is often called a 'model', and 'model' and 'hypothesis' get used interchangeably with 'theory'. A 'theoretical model' is a hypothesis describing the components and interactions thought sufficient to explain the behaviour of the target.

In the philosophy of science, the term 'model' is mostly applied to the 'source': a pre-existing system used as an analogy in devising the hypothesis. This can be a physical system e.g. a pump as the source of hypotheses for the functioning of the heart, or a system of theory, such as mathematics. If the target is to engineer better robots, then biology can be a source of ideas. On the other hand, concepts from robotics can also be a source of theory for biology, e.g. in biomechanics.

However, in this chapter we will use another meaning of 'model': the function labelled, in figure 1, as 'simulation'. Simulation (interpreted widely) involves representing the hypothesis in a specific instantiation - as a set of equations, or a computer program, or as a more obviously physical system such as a scale model. The important feature, compared to a verbal hypothesis, is that a simulation can be 'run' to automatically derive the consequences of the hypotheses it embodies.

For the following discussion, I wish to focus on how "robots can be used as physical models of animals to address specific biological questions" p.777 (Beer et al., 1998). Unlike computational models, this requires the system to have unmediated contact with a real external environment rather than a simulated environment. Unlike much of 'biologically-inspired' robotics, it requires that the aim in building the system should be to address a biological hypothesis or demonstrate understanding of a biological system. A wide range of such work is reviewed in Webb (2001). The results to date can be summarised in three broad categories:

1. The robot implementation has served to confirm the adequacy of a hypothesis, when implemented in a real system, to account for the particular behaviour. An example is the reproduction of ant homing behaviour by a robot (Möller et al., 1998) based on the 'snapshot' model of landmark navigation (Cartwright and Collett, 1983). Interestingly this work has also led to the proposal and successful testing of several simplified versions of the original hypothesis (Lambrinos et al., 2000).
2. The robot has shown that, when placed in a sufficiently realistic interaction with the environment, some simple control hypotheses suffice to explain surprisingly complex behaviours. An example is the demonstration of the sufficiency of particular optical motion cues to control tasks such as obstacle avoidance and altitude control in flying insects (Franceschini et al., 1992; Srinivasan and Venkatesh, 1997).
3. The robot building process has revealed the hypothesis to be inadequate - and has suggested some critical issues that need to be addressed. An example is the work on cockroach walking control in which a realistic physical replica of cockroach mechanics has prompted closer investigation of differences in the movement of front, middle and hind legs (Quinn and Ritzmann, 1998).

But why use robots to simulate animals? How does this methodology differ from alternative approaches to modelling in biology? To answer these questions it is necessary to understand the different ways in which models can vary.

2 Dimensions for describing models

Figure 2 presents a seven-dimensional 'space' of possible biological models. The origin is taken to be using the system itself as its own model, to cover the view expressed by Rosenblueth and Wiener (1945) as "the best material model of a cat is another, or preferably the same, cat". A model may be distanced from its target in terms of abstraction, approximation, generality, relevance, levels of organisation, material basis, or match to the target's behaviour. Exactly what is meant here by each of these dimensions, and in what ways they are (and are not) related will be discussed in detail in what follows.

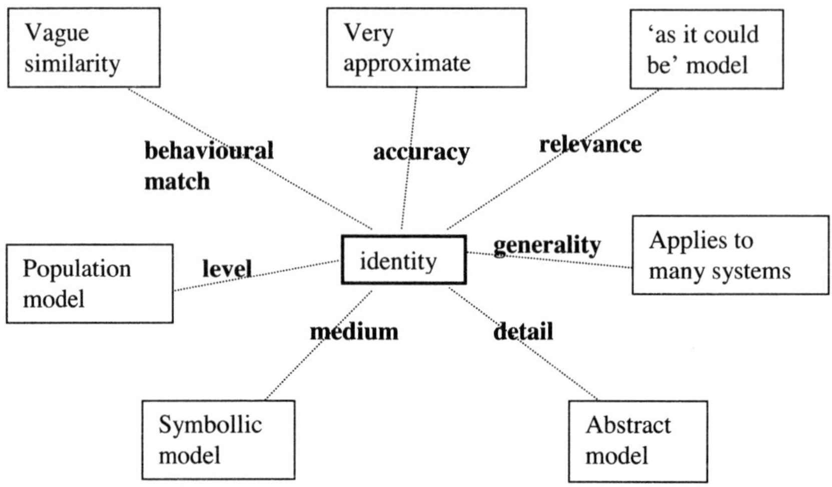

Figure 2. Dimensions of modelling

2.1 Relevance

Models can differ in the extent to which they are intended to represent, and to address questions about, some real biological system. For example Huber and Bulthoff (1998) use a robot in an investigation of the hypothesis that a single motion-sensitive circuit can control stabilisation, fixation and approach in the fly. This work is more directly applicable to biology than the robot work described in Srinivasan et al. (1999) utilising bee-inspired methods of control from visual flow-fields, which does not principally aim to answer questions about the bee.

The main issue for relevance is the ability of the model to generate testable hypotheses about the biological system it is drawn from. Are the mechanisms in the model explicitly mapped back to processes in the animal, as hypotheses about its function? This could concern neural circuitry, e.g. in a robot model of auditory localisation of the owl (Rucci and Edelman, 1999). But it can also occur at a relatively high level, such as using shaping methods in robot learning (Saksida et al., 1997) or involve testing a simple algorithm on a robot such as the use of sky-polarisation as a compass in ant navigation (Lambrinos et al., 1997). Or it may relate to the physics of the animal environment interaction, for example the suggestion from the robot studies of Triantafyllou and Triantafyllou (1995) that fish use the creation of vortexes as a means of efficient tail-fin propulsion.

Relevance is what at least some authors mean by the term 'realism' in describing models; however others define 'realism' as the amount of detail the model includes (see below). It is important to appreciate that these are separate issues, that is, models do not necessarily have to be detailed to be relevant to biology.

2.2 Level

This dimension concerns the hierarchy of physical/processing levels that a given biological model could attempt to represent. Any hypothesis will usually have 'elemental units' whose properties are assumed rather than subject to further explanation. In biology these can range from the lowest known mechanisms such as the physics of chemical interactions through molecular and channel properties, membrane dynamics, compartmental properties, synaptic and neural properties, networks and maps, systems, brains and bodies, perceptual and cognitive processes, up to social and population processes (Schwartz, 1990). In robot models, the level can vary from rule-based, such as the decision mechanisms involved in the collective sorting behaviour of robot 'ants' (Holland and Melhuish, 1999), through high level models of brain function such as the control of eye movements (Shibata and Schaal, 1999), to models of specific neuron connectivity hypothesised to underlie the behaviour, such as identified neural circuitry in the cricket (Webb and Scutt, 2000), and even to the level of dendritic tree structure that explains the output of particular neurons such as the 'looming' detector found in the locust (Blanchard et al., 1999).

It is a misconception to assume that the level of a model determines its biological relevance. A model is not necessarily made to say more about biology just by including lower-level mechanisms e.g. most 'neural network' controlled robots have little to do with understanding biology, whereas, as described above, some robots using quite high-level algorithms explicitly address biological issues. In general, if the aim of the work is to be relevant to biology, the level of mechanism modelled by the robot will reflect closely the level of information currently available from biology. It is also important to distinguish level from accuracy (see below) as it is quite possible to inaccurately represent any level.

2.3 Generality

A more general model is one that applies to a greater number and/or variety of real systems. Ayers et al. (1998) claim "locomotory and taxis behaviours of animals are controlled by mechanisms that are conserved throughout the animal kingdom" and thus their model of central pattern generators is taken to be of high generality. Strictly speaking, a model must be relevant to be general - if it doesn't apply to any specific system, then how can it apply to many systems (Onstad, 1988)? Of course a model does not have to be general to be relevant to some specific system. The obvious way to demonstrate that a model is general is to show how well it succeeds in representing a variety of different specific systems. For many models labelled 'general' this doesn't happen. When it is attempted, it usually requires a large number of extra situation or task specific assumptions to actually get data from the model to compare to the observed target. It is not always clear that this any better than building specific models in the first place.

The most common confusion regarding generality is that what is abstract will therefore be general. This is often claimed in Artificial Life research for example, but does not necessarily follow - a highly simplified description might apply to no particular animal system, rather than to all systems. This raises the important point that "generality has to be found, it cannot simply be declared" p.155 (Weiner, 1995). That is to say the generality of a model depends on the true nature of the target(s). If different animals function in different ways then trying to generalise over them won't work - you are left studying an empty set. Biology has often found that the discovery and elucidation of general mechanisms tends to come most effectively from close exploration of well-chosen specific instantiations such as the squid giant axon.

2.4 Abstraction

Abstraction concerns the number and complexity of mechanisms included in the model; a more detailed model is less abstract. The 'brachiator' robot models studied by Saito and Fukuda (1996) illustrate different points on this spectrum: their initial model was a fairly simple two-link device, but in more recent work they argue that it is important to base the robot body on exact measurements from the simiang. Abstraction is not just a measure of the simplicity/complexity of the model however (Brooks and Tobias, 1996) but is relative to the complexity of the target. Thus a simple target might be represented by a simple, but not abstract, model, and a complex model still be an abstraction of a very complex target.

Some degree of abstraction is bound to occur in most model building, and it is sometimes taken as a defining characteristic of the modelling process. Although increasing detail often involves moving to a lower level of description, is important to note that there can be very abstract models of low-level processes - such as genetic algorithms - and very detailed models of high-level processes - such as the population dynamics of a particular ecological system.

How much abstraction is considered appropriate seems to largely reflect the taste of the modeller for simple, elegant models or for closely detailed system descriptions. Complex models can be harder to implement, understand, replicate or communicate; simpler models reduce the risk of merely data-fitting, by having fewer free parameters. However abstraction carries risks. The existence of an attractive formalism might end up imposing its structure on the problem so that alternative, possibly better, interpretations are missed. Details abstracted away might turn out to actually be critical to understanding the system. As Kaplan (1964) notes the issue is often not just 'over-simplification' per se, but whether we have "simplified in the wrong way" or that "what was neglected is something important for the purposes of that very model".

2.5 Accuracy

Accuracy is an assessment of how well the mechanisms in the model reflect the real mechanisms in the target. Hannaford et al. (1995) lay out their aims in building a robot replica of the human arm as follows: "Although it is impossible to achieve complete accuracy, we attempt to base every specification of the system's function and performance on uncontroversial physiological data". The term 'accuracy' is thus being used to refer to structural validity: "if it not only produces the observed real system behaviour but truly reflects the way in which the real system operates to produce this behaviour" Zeigler (1976). Anti-realists argue that we can never know with certainty that a scientific model truly reflects the real system's workings (Oreskes et al., 1994). Nevertheless it is possible to discuss, relative to the contemporary scientific context in which the model is built, the extent to which the mechanisms included seem to agree or conflict with what we believe about the workings. For example, would we expect to be able to find independent experimental verification for the internal mechanisms that the model includes?

Accuracy can be distinguished from relevance: it is possible for a model to address real biological questions without utilising accurate mechanisms. For example, the real task faced by the animal may be clarified by trying to get any robot system to perform comparably. Accuracy is also not synonymous with amount of detail included in the model. One can't assume a model with lots of complex detail is accurate, without actually checking that the details are correct, and some abstractions might prove more accurate that others.

2.6 Match

This dimension concerns how the model will be assessed. In one sense it concerns testability: can we potentially falsify the model by comparing its behaviour to the target? For example, the possibility that the lobster uses instantaneous differences in concentration gradients between its two antennules to do chemotaxis was ruled out by testing a robot implementation of this algorithm in the real lobster's flow-tank (Grasso et al., 2000). But there is still much variability in the possible match between the behaviours. Are the behaviours indistinguishable or merely similar? Are informal, expert or systematic statistical investigations to be used as criteria for assessing similarity? Is a qualitative or quantitative match expected? Can the model both reproduce past data and predict future data? Some modelling studies provide little more than statements that, for example, "the overall behaviour looked quite similar to that of a real moth" p.375 (Kuwana and Shimoyama, 1995). Others make more direct assessment, e.g. Harrison and Koch (1999) have tested their analog VLSI optomotor system in a real fly flight simulator and "repeated an experiment often performed on flies" to assess directly how much detail of the fly's behaviour they can reproduce.

There are inevitable difficulties in drawing strong conclusions about biological systems from the results of robot models. As with any model, the performance of similar behaviour is never sufficient to prove the similarity of mechanisms. If the model doesn't match the target then we can reject the hypothesis that led to the model or at least know we need to improve our model. If it does match the target, better than any alternatives, then the hypothesis is plausible to the extent that we think it unlikely that such similar behaviour could result from completely different causes. However carrying out the comparison of model and target behaviours can be a sufficiently complex process that neither of these arguments can be completely relied upon. The intepretation of the model behaviour might be flexible enough to account for any measurements, particularly by judicious parameter tuning or ad hoc introduction of small modifications.

2.7 Medium

Hypotheses can be instantiated as models in various different forms, and hardware implementation is one of the most striking features of robot models. Kuwana and Shimoyama (1995) use actual biological sensors - the antennae of moths - on their robot model and note these are 10,000 times more sensitive than available gas sensors. More commonly, models share some physical properties with biology e.g. gas sensing is substituted for pheromone sensing in Ishida et al. (1999) robot model of the moth, but they replicate other physical features of the sensor, such as the way that the moths wings act as a fan to draw air over the sensors. Or modelling may substitute similar physical properties e.g. the robot model of chemotaxis in C. Elegans (Morse et al., 1998) uses a light source as an analog for a chemical gradient, while preserving a similar sensor layout and sensitivity. Sometimes models use quite different properties to stand in for the properties specified in the target e.g. the use of different coloured blocks in a robot arena to represent 'food' and 'poison'.

What is important are the constraints the medium imposes on the operation of the model. The medium may contribute directly to the accuracy and relevance of the model, or simply make it easier to implement, run or evaluate as described by Quinn and Espenscheid (1993): "Even in the most exhaustive [computer] simulations some potentially important effects may be neglected, overlooked or improperly modelled. It is often not reasonable to attempt to account for the complexity and unpredictability of the real world. Hence implementation in hardware is often a more straightforward and accurate approach for rigorously testing models of nervous systems"

3 Conclusion

As this analysis of dimensions has shown, there are many decisions to be made when building a model. There is not a single correct way to implement hypotheses, but rather a range of considerations and constraints that need to be taken into account. The research I have referred to has in common two main characteristics: an interest in building simulations that are relevant representations of real biological hypotheses; and embodying these simulations in robot form. But on many of the other dimensions there is a great deal of variety: in the level of organisation represented; in the generality of the problem addressed; in the amount of detail included; in the accuracy with which biological mechanisms have been copied; and in the means by which the results have been compared to the biological system.

Using a robot as a model has the advantage that it tends to force integration - across levels, and mechanisms. This is because the aim is to build a complete system that connects action and sensing to achieve a task in an environment. This may limit the individual accuracy of particular parts of the model because of necessary substitutions, interpolations and so on, but this is considered a price worth paying for the benefits of gaining a more integrated understanding of the system and its context, in particular the "tight interdependency between sensory and motor processing" (Pichon et al., 1989). Physical implementation imposes the necessity that that all parts of the system function together and produce a real output.

For example, Hannaford et al. (1995) argue that "Physical modelling as opposed to computer simulation is used to enforce self consistency among co-ordinate systems, units and kinematic constraints" in their robot arm. Another important consideration is that using identity in parts of a model can sometimes increase accuracy at relatively little cost. Using real water or air-borne plumes, or real antennae sensors, saves effort in modelling and makes validation more straightforward.

However a more fundamental argument for using physical models is that an essential part of the problem of understanding behaviour is understanding the environmental conditions under which it must be performed - the opportunities and constraints that it offers. If we simulate these conditions, then we include only what we already assume to be relevant, and moreover represent it in a way that is inevitably shaped by our assumptions about how the biological mechanism works. Thus Flynn and Brooks (1989) argue that "unless you design, build, experiment and test in the real world in a tight loop, you can spend a lot of time on the wrong problems".

Bibliography

J. Ayers, P. Zavracky, N. Mcgruer, D. Massa, V. Vorus, R. Mukherjee, and S. Currie. A modular behavioural-based architecture for biomimetic autonomous underwater robots. In *Autonomous Vehicles in Mine Countermeasures Symposium*, 1998.

R.D. Beer, H.J. Chiel, R.D. Quinn, and R.E. Ritzmann. Biorobotic approaches to the study of motor systems. *Current Opinion in Neurobiology*, 8(6):777–782, 1998.

M. Blanchard, P.F.M.J Verschure, and F. Claire Rind. Using a mobile robot to study locust collision avoidance responses. *International Journal of Neural Systems*, 9(5):405–410, 1999.

V. Braitenberg. *Vehicles: experiments in synthetic psychology.* MIT Press, Cambridge, MA, 1984.

R. J. Brooks and A. M. Tobias. Choosing the best model: level of detail, complexity and model performance. *Mathematical Computer Modelling*, 24(4):1–14, 1996.

B. Cartwright and T. Collett. Landmark learning in bees. *Journal of Comparative Physiology A*, 151:521–543, 1983.

K.H. Chan and P.M. Tidwell. The reality of artificial life: can computer simulations become realizations? In *submission to Third International Conference on Artificial Life*, 1993.

A.M. Flynn and R.A. Brooks. Battling reality. Technical Report A.I. Memo 1148 M.I.T. A.I. Lab, M.I.T., 1989.

N. Franceschini, J.M. Pichon, and C. Blanes. From insect vision to robot vision. *Philosophical Transactions of the Royal Society B*, 337:283–294, 1992.

F. Grasso, T. Consi, D. Mountain, and J. Atema. Biomimetic robot lobster performs chemo-orientation in turbulence using a pair of spatially separated sensors: Progress and challenges. *Robotics and Autonomous Systems*, 30(1-2):115–131, 2000.

B. Hannaford, J. Winters, C-P Chou, and P-H Marbot. The anthroform biorobotic arm: a system for the study of spinal circuits. *Annals of Biomedical Engineering*, 23:399–408, 1995.

R. R. Harrison and C. Koch. A robust analog VLSI motion sensor based on the visual system of the fly. *Autonomous Robotics*, 7(3):211–224, 1999.

O. Holland and C. Melhuish. Stigmergy, self-organization and sorting in collective robotics. *Artificial Life*, 5:173–202, 1999.

S.A. Huber and H.H. Bulthoff. Simulation and robot implementation of visual orientation behaviour of flies. In R. Pfeifer, B. Blumberg, J.A. Meyer, and S.W. Wilson, editors, *From animals to animats 5*, pages 77–85, Cambridge, Mass., 1998. MIT Press.

H. Ishida, A. Kobayashi, T. Nakamoto, and T. Moriisumi. Three dimensional odor compass. *IEEE Transactions on Robotics and Automation*, 15:251–257, 1999.

A. Kaplan. *The conduct of enquiry*. Chandler, San Francisco, 1964.

Y Kuwana and H Shimoyama, I;Miura. Steering control of a mobile robot using insect antennae. In *IEEE International Conference on Intelligent Robots and Systems*, volume 2, pages 530–535, 1995.

D. Lambrinos, M. Maris, H. Kobayashi, T. Labhart, R. Pfeifer, and R. Wehner. An autonomous agent navigating with a polarized light compass. *Adaptive Behaviour*, 6(1):175–206, 1997.

D. Lambrinos, R. Möller, T. Labhart, R. Pfeifer, and R. Wehner. A mobile robot employing insect strategies for navigation. *Robotics and Autonomous Systems*, 30(1-2):39–64, 2000.

R. Möller, D. Lambrinos, R. Pfeifer, T. Labhart, and R. Wehner. Modeling ant navigation with an autonomous agent. In R. Pfeifer, B. Blumberg, J.A. Meyer, and S.W. Wilson, editors, *From animals to animats 5*, Cambridge, Mass., 1998. MIT Press.

T.M. Morse, T.C. Ferree, and S.R. Lockery. Robust spatial navigation in a robot inspired by chemotaxis in *Caenorrhabditis elegans*. *Adaptive Behaviour*, 6(3/4):393–410, 1998.

D.W. Onstad. Population-dynamics theory - the roles of analytical, simulation, and supercomputer models. *Ecological Modelling*, 43(1-2):111–124, 1988.

N. Oreskes, K. Shrader-Frechette, and K. Belitz. Verification, validation and confirmation of numerical models in the earth sciences. *Science*, 263:641–646, 1994.

J-M Pichon, C. Blanes, and N. Franceschini. Visual guidance of a mobile robot equipped with a network of self-motion sensors. In W.J. Wolfe and W.H. Chun, editors, *Mobile Robots IV*, volume 1195, pages 44–53, Bellingham, 1989. Society of Photo-optical Instrumentation Engineers.

R.D. Quinn and K.S. Espenscheid. Control of a hexapod robot using a biologically inspired neural network. In R.D. Beer, R.E. Ritzmann, and T. McKenna, editors, *Biological Neural Networks in Invertebrate Neuroethology and Robotics.* Academic Press, London, 1993.

R.D. Quinn and R.E. Ritzmann. Construction of a hexapod robot with cockroach kinematics benefits both robotics and biology. *Connection Science*, 10:239–254, 1998.

A. Rosenblueth and N. Wiener. The role of models in science. *Philosophy of Science*, 12(4):316–321, 1945.

M Rucci and J Edelman, GM;Wray. Adaptation of orienting behavior: from the barn owl to a robotic system. *IEEE Transactions on Robotics and Automation*, 15(1):p 96–110, February 1999.

F. Saito and T. Fukuda. A first result of the brachiator III - A new brachiation robot modeled on a siamang. In C. Langton and K. Shimohara, editors, *Proceedings of ALife V*, Cambridge MA, 1996. MIT Press.

L.M. Saksida, S.M. Raymond, and D.S. Touretzky. Shaping robot behavior using principles from instrumental conditioning. *Robotics and Autonomous Systems*, 22(3-4):231–249, 1997.

E.L. Schwartz. Introduction. In E.L. Schwartz, editor, *Computational Neuroscience.* MIT Press, Cambridge, Mass., 1990.

T. Shibata and S. Schaal. Robot gaze stabilisation based on mimesis of oculomotor dynamics and vestibulocerebellar learning. *Advanced Robotics*, 13(3):351–352, 1999.

M.V. Srinivasan and S. Venkatesh. *From Living Eyes to Seeing Machines.* Oxford University Press, Oxford, 1997.

M.V. Srinivasan, J.S. Chahl, K. Weber, and S. Venkatesh. Robot navigation inspired by principles of insect vision. *Robotics and Autonomous Systems*, 26:203–216, 1999.

M.S. Triantafyllou and G.S. Triantafyllou. An efficient swimming machine. *Scientific American*, 272(March):40–48, 1995.

B. Webb. Can robots make good models of biological behaviour? *Behavioural and Brain Sciences*, 24(6):1033–1094, 2001.

B. Webb and T. Scutt. A simple latency dependent spiking neuron model of cricket phonotaxis. *Biological Cybernetics*, 82(3):247–269, 2000.

J. Weiner. On the practice of ecology. *Journal of Ecology*, 83(1):153–158, 1995.

B. P. Zeigler. *Theory of Modelling and Simulation.* John Wiley, New York, 1976.

Part II

From sensing toward perception

Spiking neuron controllers for a sound localising robot

Barbara Webb

Institute of Perception, Action and Behaviour
School of Informatics, University of Edinburgh

Abstract Female crickets can find a mate by recognising and walking or flying towards the male calling song. Using a robot to model this behaviour, we explore of the functionality of identified neurons in the insect, including the roles of multiple sensory fibres, mutually inhibitory connections, and brain neurons with pattern-filtering properties.

1 Introduction

We are interested in the complexity of behaviour that can result from the interaction of a small number of neural elements. We have chosen to focus on modelling the specific neural circuit underlying a particular animal behaviour, the sound localisation ability of the cricket. We consider the modelling of this 'simple' insect behaviour to be the focal point for investigating a range of interesting, interconnected general issues for biology. These include: the functional significance of low-level neural properties; the importance of physical embodiment as a solution and constraint on behaviour; the possible connections between biological solutions and conventional engineering approaches to sensorimotor control and integration; and how we can scale up from single behaviours to explain the flexible interaction of multiple adaptive systems within an organism.

Female crickets can locate conspecific males by moving towards the species-specific calling song the males produce by opening and closing their wings. Typical male songs consist of groups of short sound bursts (e.g. for *Gryllus bimaculatus*, four 20 ms 'syllables' of 4.7 kHz sound make up a 'chirp') produced several times a second. Females appear to be particularly selective for the repetition rate of syllables within each chirp. The neuroethology of this system has been extensively studied (reviewed in Pollack, 1998) and we have built a series of robot models that have led to interesting reinterpretations of this data. In Webb (1995) and Lund et al. (1997, 1998) it was shown that the recognition and localisation aspects of the task could be closely interlinked, allowing a surprisingly simple controller to produce the same kind of selective approach behaviour in a robot as was observed in the cricket. In Webb and Scutt (2000) robot hardware was interfaced with a spiking neural network simulation, and it was shown that the algorithmic controller explored previously could be captured in a four neuron circuit (figure 1). Although the essence of this circuit is a Braitenberg-like connection between the input on each side and the motor outputs (Braitenberg, 1984), the actual function is more subtle, it exploits spike timing and dynamic synaptic properties to be selective for the temporal pattern of signals in a manner resembling the female cricket.

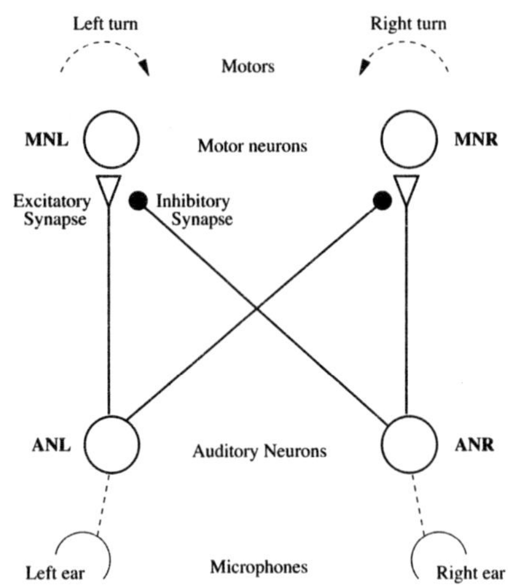

Figure 1. The four neuron circuit used in previous work (Webb and Scutt, 2000). The auditory neurons (AN) represent the first stage of processing in the cricket's prothoracic ganglion in which a pair of neurons receives direct input from the auditory nerve. The motor neurons (MN) produce an output signal to turn left or right depending on which AN neuron fires first. The AN-MN synapses exhibit depression, so the response is best for a signal with appropriate temporal patterning.

Effective biological modelling requires repeated cycles of implementation, testing and refinement. The neural circuit shown in figure 1 could reproduce much of the cricket behaviour, but this is not sufficient to conclude that the cricket's neural circuit is the same. There may be alternative wirings that work as well or better, or are more consistent with the anatomical and physiological data. For example, the inhibitory cross-connections from prothoracic auditory neurons (AN) to opposite motor neurons (MN) in the model were not particularly plausible, and may not be necessary to create a 'winner-take-all' response to the stronger auditory signal. Instead the AN or MN could be mutually inhibitory. There is neurophysiological evidence from the cricket of mutual inhibition at the prothoracic AN level, mediated by so-called omega neurons (ON). In addition, the cricket is unlikely to have such a direct connection between auditory input and motor output. There is information about intermediate stages of processing in brain neurons from Schildberger (1984) that can be more directly addressed. Finally, the underlying model of neural activity used in Webb and Scutt (2000) was itself somewhat abitrary; so a representation of spiking dynamics that links more closely to current theoretical neuroscience could improve the biological plausibility of the circuits implemented on the robot.

2 Methods

The robot base used in the following experiments is the 'Koala' (K-Team S.A.). It has a six-wheeled base measuring approximately 40x30x20 cm with two drive motors. Although the actuation bears little direct resemblance to the cricket it is possible to use this robot to replicate appropriate speeds and turning rates. This robot has a Motorola 68331 processor, programmable in C, which was used to transfer sensor data and motor commands between the robot and the neural simulation described below.

2.1 Sensors

The robot has a customised auditory processing circuit (Lund et al., 1997) based on the pressure-difference receiver instantiated by the cricket's peripheral auditory system (Michelsen et al., 1994). The cricket has its eardrums on its front legs, and the two sides are connected by an air-filled tracheal tube. Thus the vibration of each ear drum is a result of both the sound reaching it externally and the sound transferred internally. The relative phase between these sounds will depend on the direction of the sound, its wavelength, and the length of the tracheal tube. For a particular wavelength, the phase-cancellation occuring at the tympanum will systematically alter the amplitude of vibration for sound sources from different directions.

For the robot, two microphones separated by 18mm (1/4 wavelength of the carrier frequency - 4.7kHz - of cricket song) receive and amplify the sound. The signal from the left microphone is delayed by 53 microseconds

(1/4 the phase of 4.7kHz) and then subtracted from the right; and vice-versa. The amplitude of the resulting waves is thus increased for sound on the same side as the microphone and decreased on the opposite side. The amplitude is measured using a RMS circuit, resulting in two analog signals representing the amplitude of vibration of the tympani in the cricket. These signals are converted on the robot into Poisson distributed spike trains proportional to the amplitude, and used as synaptic inputs to the neural model.

2.2 Neural model

The neural simulator is a C++ program running on a PC under Linux. Each neuron in the circuit is represented by a single compartment 'leaky integrate and fire' model, based on the RC circuit description of neurons in Koch (1999). Their state is a representation of the potential difference across the neural membrane (the membrane potential). They have a base potential, to which they will decay in the absence of external input. Synapses attach to them and raise and lower their potential when their pre-synaptic neurons spike. If the potential rises above a specific level, the neuron will fire, sending a spike to any connected output synapses; then the potential will reset to some lower value, and the neuron will enter a refractory period when it will be unable to receive synaptic input.

The model synapses are conductance based, which is to say they model the conductance and battery potential of the ion channels which open when a synapse is activated, pulling the membrane potential towards the battery potential with a strength proportional to the conductance. This is more biologically realistic than the standard 'charge-dump' or current injection synapses used in most artificial neural networks, which directly raise or lower the membrane potential by depositing a small or large packet of charge into the neuron. The conductance decays exponentially to zero, but receives a boost when a spike arrives from the input neuron. The size of the boost to the conductance is effectively the 'weight' of the connections. The weight is further affected by short-term facilitation and depression mechanisms. There are also variable synaptic delays in the time from the firing of the presynaptic neuron to the reception of the excitation or inhibition by the post synaptic neuron.

2.3 Experimental paradigm

The neural simulation system described above is embedded as part of a C++ program for running robot experiments. This system deals with the serial transfer of sensory data and motor data between the robot and the neural simulation, and is designed for easy transfer between different robot bases and the addition of arbitrary new sensor inputs or motor outputs. It can also be run without the robot, using previously recorded or artificially generated data files as input, which is particularly useful when tuning circuit parameters. It automatically incorporates data from an overhead camera tracking system into the data record produced from running the robot.

The results described below were produced in two ways: analytical and simulation results; and results from running the robot. The latter experiments occurred in the normal lab environment, within a 2x1.6 metre space determined by the overhead camera's field of view. The sound was a computer-generated song. The standard song consists of 20 ms bursts (syllables) of 4.7 kHz sound, with 20 ms gaps between them, grouped into a four-burst chirp followed by 340 ms silence. This is the characteristic calling song of male *Gryllus bimaculatus*. In some experiments the pattern was varied to produce different 'syllable repetition intervals': the length of the chirp was held constant at 250 ms, and the length of syllables within the chirp varied from 5 ms to 45 ms, with intersyllable gaps of the same length as the the syllable.

3 Auditory processing

The previous neural model tested on the robot focussed on one pair of identified ascending interneurons (AN1) in the cricket's prothoracic ganglion that appear to be critical for phonotaxis. AN1 respond best to sound at the calling song carrier frequency, and clearly encode the pattern of the song in their spiking response. Hyperpolarising one of the pair leads to a change in walking direction (Schildberger and Horner, 1988). In the current model we incorporate more of the neural data from the cricket, motivated by functional hypotheses about the various neural elements, aiming to reproduce specific response properties. The new model is illustrated in figure 2.

First, we have represented the parallel sensory fibres that provide input from the eardrums via the leg nerve to the prothoracic ganglion. In the cricket there are around 50–60 sensory neurons, with perhaps half of these tuned to the calling song frequency (Esch et al., 1980). There is some evidence of range fractionation for different sound amplitudes. Recordings from the leg nerve show neural time constants in the order of a millisecond and thus clear copying of time varying patterns, with some adaptation within the first few milliseconds of response to a sound burst. Our simulated auditory nerve consists of eight parallel fibres and the response has been tuned to resemble the cricket data.

Second, we have included a second pair of neurons that receive input

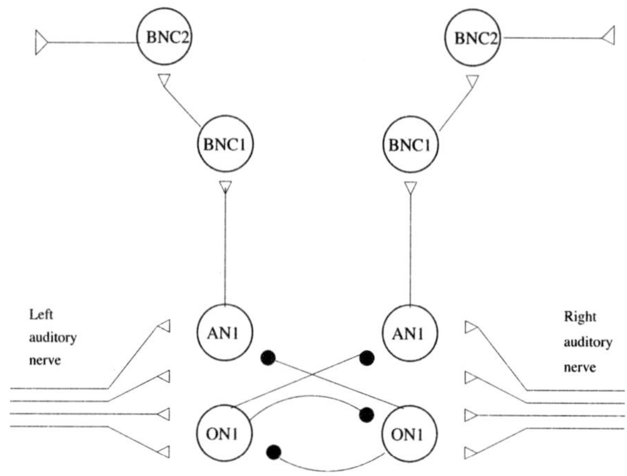

Figure 2. The new neural circuit for phonotaxis, based on cricket neuroanatomy, as described in text. 'AN', 'ON' and 'BN' refer to specific 'ascending', mutually connected 'omega', and 'brain' neurons that have been identified in the cricket nervous system.

from the auditory nerve, based on the omega neurons (ON) described in the cricket (Wohlers and Huber, 1981). These are mutually inhibitory and also inhibit the opposite ascending interneurons. The most obvious function of these connections is to increase the difference in activation between the two sides, to emphasise the directionality of the response. Elegant experiments from Horseman and Huber (1994) have shown that inhibition from the opposite side makes a difference of several spikes per chirp to the AN1 response. Figure 3 illustrates the replication of this effect in our model (compare to their figure 3). Here the decrease in firing rate to an ipsilateral sound level is proportional to the contralateral sound level. Although the decrease in firing rate is not large, one advantage may be gain control. For varying amplitudes of sound (experienced as the cricket or robot approaches the sound source) the amplitude of the response adapts, because the inhibitory signal also changes. This allows the circuit to encode the relevant difference between the ears within a similar range of firing rates, without saturation.

Third, we have included a stage of neural processing in the 'brain' before signals are sent to the motor output. The most comprehensive study of the role of cricket brain neurons in phonotaxis is provided by Schildberger

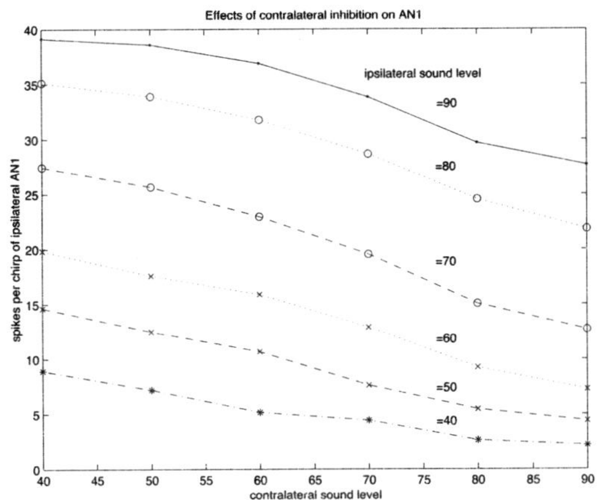

Figure 3. Effect of inhibitory cross connections on the firing rate of AN1. The ipsilateral firing rate decreases as the contralateral sound level increases.

(1984), who suggests a possible filtering circuit for syllable rate recognition by the female. He identifies two main classes of auditory responsive cells: BNC1 which appears to get direct input from AN1; and BNC2 which get input via BNC1. These neurons vary in their response to the sound pattern, appearing to act as low-pass, highpass and bandpass filters for the syllable rate. We attempted to directly copy these response patterns in our simulated neurons, but there were several difficulties. In particular, it seemed impossible distiguish very long syllables from very rapid syllables (in order to have a highpass filter) as they both produce the same response – continuous firing – in the AN1 neurons.

After some systematic searching of the parameter space we found that a lowpass filter could be produced by having a depressing synaptic connection from AN1 to BN1. BN1 would thus fire several spikes at the onset of syllables. If the syllable rate was fast, the depression would not recover and fewer spikes would occur. At slower rates, the recovery time between syllables was longer and the chance of spiking at the next syllable onset increased; however as the time between onsets increased the overall firing rate started to decrease. A second depressing synaptic connection from BN1 to BN2, with a relatively slow time constant, performed temporal summation of the BN1 output so that these slower responses were not usually sufficient to produce a spike, making BN2 a bandpass filter for the correct syllable rate. The ideal syllable rate is thus the fastest rate for which clear gaps for recovery occur in the AN1 response. In figure 4 we show the response of our BN1 and BN2 neurons to song patterns identical to those used by Schildberger, i.e. using equal length chirps with syllable repetition intervals (SRI) ranging from 10 ms to 90 ms. It can be seen that BN1 has a moderately bandpass response which is sharpened by BN2. The response of BN2 is very similar to the BN2a neuron reported by Schildberger. In the top plots which show results with simulated sound inputs, the apparent best tuning is to an SRI of 26 ms. However, using the same neural parameters with real sound input, as shown in the lower plots, syllables at this rate are not so clearly coded, and the best response in BN2 moves to an SRI of between 42 ms and 58 ms.

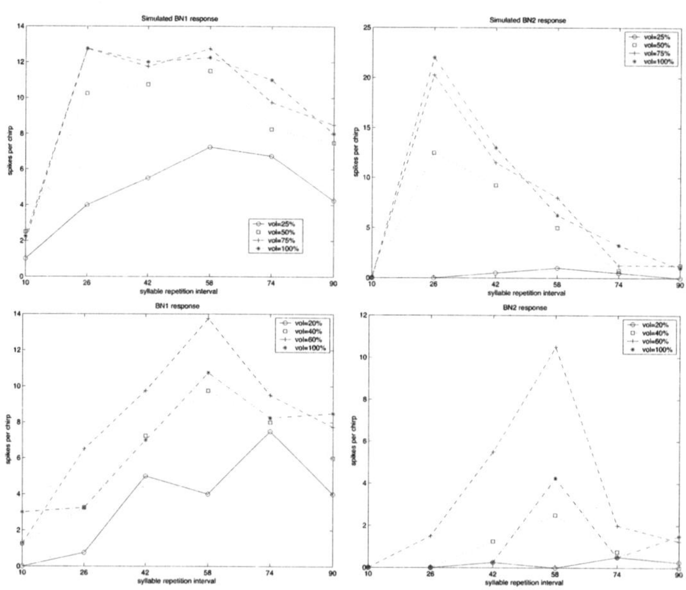

Figure 4. Spike rates of simulated brain neurons to songs with different syllable repetition intervals. Upper plots, simulated inputs; lower plots, real sound inputs. Left, the BN1 neuron shows a lowpass/bandpass response; right, the BN2 neuron shows a bandpass response.

4 Motor response

The BN2 neurons respond best to clear firing patterns in AN1, which generally correspond to the louder side. We found it was not necessary to include any explicit mechanism for comparing the firing rates or latencies at the BN2 level. We could simply take a spike in the left or right BN2 as indicating the need to turn in that direction to approach the sound.

Our aim in designing the motor circuit was to reproduce the specific kinds of moves, turns and stops observed in crickets performing phonotaxis, with as few neural connections as possible. Current models of motor control for six-legged walking in insects, see Cruse et al. (1998), suggest several features to include in our circuit. Motor patterns tend to be self-sustaining through local feedback or Central Pattern Generator (CPG) activity, so that only a trigger signal is required to start the movement. Steering appears to be modulated by fairly simple turn signals from the brain interacting with this pattern generator so as to modify limb movements appropriately. The neural circuitry we have used is adapted from the scheme devised by Chapman (2001) for a robot model of cricket escape behaviour. It is illustrated in figure 5. The paired burst generators (BG), when initiated by an incoming spike, mutually activate each other to produce a continuous burst of spikes that go to right and left forward neurons (RF and LF) and drive the robot forward. The length of the burst is limited by the eventual activation of the STOP neuron which inhibits the BG neurons. One trigger for movement is a spike in the left or right BNC2. These act via a right or left turn neuron (RT or LT) to additionally modulate the forward velocity by appropriate excitatory and inhibitory connections to RF and LF.

Figure 6 shows the performance of the robot using the combined phonotaxis and motor circuits to track cricket song. Thirty trials were run, from 3 different starting positions: directly in front of the speaker, or starting with the speaker to the right or to the left. It can be seen that the robot is capable of successfully locating the sound source, producing a cricket-like path that zigzags towards the sound.

5 Conclusions

Our approach to biologically-based robotics assumes that to exploit biological designs we need to understand how biological systems work. Thus we have opted for close study of a specific system, and aimed to match the known behaviour and physiology in some detail. The neural model used in this work has a number of characteristics that we believe are important in understanding the flexibility and adaptability of animal behaviour. In

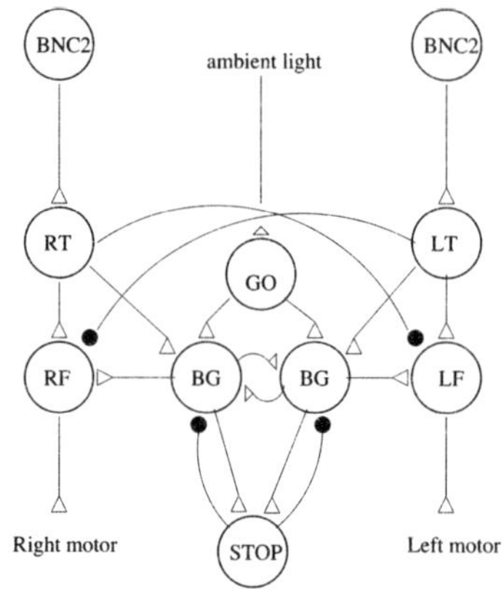

Figure 5. Neural circuit for motor control. The BG pair form a burst generator that excites right forward (RF) and left forward (LF) neurons when triggered by RT, LT or GO till deactivated by STOP. BNC2 is the output from the auditory circuit in figure 2 and RT and LT produce right and left turns respectively by excitatory and inhibitory modulation of RF and LF.

particular we represent many of the dynamics of synaptic activity rather than use a simple 'weight' model as found in conventional artificial neural nets. The result is that it is easier to implement - with a small number of neurons - various important characteristics of the behaviour. This is not just a case of trading more complex individual neural models against the number of neurons needed, but that there is a natural match between the neural mechanisms and the desired behavioural characteristics. This is illustrated by the way we can tune a few temporal parameters in the neurons to produce cricket-like selectivity for the temporal pattern of the signal.

The results of our robot model also provide predictions for biological experiments. For example, we found that taking into account the range of sound levels over which the cricket performs taxis limited the possible options for filtering the temporal pattern. We also showed that the cross-inhibition occuring at the prothoracic ganglion (the omega neurons) is sufficient to enable a directional response without any further 'comparison' stage between the two sides. Our current work involves direct collaboration with researchers in cricket neurophysiology to investigate these issues.

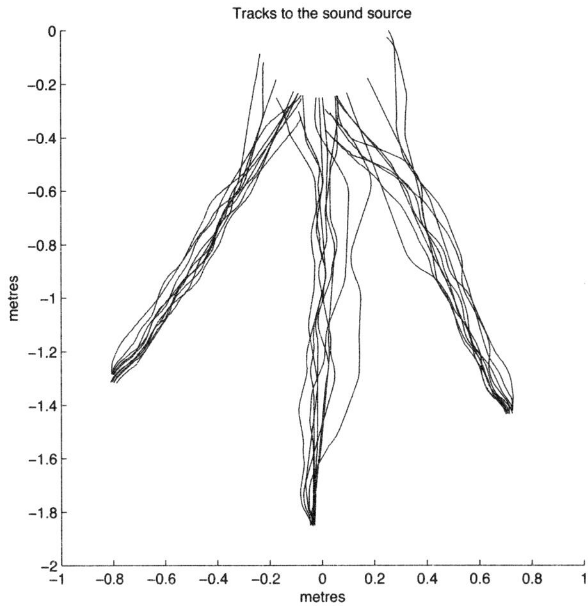

Figure 6. Tracks of the robot to a cricket song from three starting positions (thirty tracks in total). The sound source is at (0,0).

Bibliography

V. Braitenberg. *Vehicles: experiments in synthetic psychology.* MIT Press, Cambridge, MA, 1984.

T. Chapman. *Morphological and Neural Modelling of the Orthopteran Escape Response.* PhD thesis, University of Stirling, Stirling, U.K., 2001.

H. Cruse, T. Kindermann, M. Schumm, J. Dean, and J. Schmitz. Walknet – a biologically inspired network to control six-legged walking. *Neural Networks*, 11(7-8):1435–1447, 1998.

H. Esch, F. Huber, and D. W. Wohlers. Primary auditory interneurons in crickets: physiology and central projections. *Journal of Comparative Physiology*, 121:27–38, 1980.

G. Horseman and F. Huber. Sound localisation in crickets. I. Contralateral inhibition of an ascending auditory interneuron. *Journal of Comparative Physiology A*, 175:389–398, 1994.

C. Koch. *Biophysics of Computation.* Oxford University Press, 1999.

H. H. Lund, B. Webb, and J. Hallam. A robot attracted to the cricket species *Gryllus bimaculatus.* In P. Husbands and I. Harvey, editors, *Proceedings of 4th European Conference on Artificial Life*, pages 246–255. MIT Press/Bradford Books, MA., 1997.

H. H. Lund, B. Webb, and J. Hallam. Physical and temporal scaling considerations in a robot model of cricket calling song preference. *Artificial Life*, 4(1):95–107, 1998.

A. Michelsen, A. V. Popov, and B. Lewis. Physics of directional hearing in the cricket *Gryllus bimaculatus*. *Journal of Comparative Physiology A*, 175:153–164, 1994.

G. S. Pollack. Neural processing of acoustic signals. In A. N. Popper R. R. Hoy and R. R. Fay, editors, *Comparative Hearing: Insects*, pages 139–196. Springer, Berlin, 1998.

K. Schildberger. Temporal selectivity of identified auditory interneurons in the cricket brain. *Journal of Comparative Physiology*, 155:171–185, 1984.

K. Schildberger and M. Horner. The function of auditory neurons in cricket phonotaxis i influence of hyperpolarization of identified neurons on sound localisation. *Journal of Comparative Physiology A*, 163:621–631, 1988.

B. Webb. Using robots to model animals: a cricket test. *Robotics and Autonomous Systems*, 16(2-4):117–134, 1995.

B. Webb and T. Scutt. A simple latency dependent spiking neuron model of cricket phonotaxis. *Biological Cybernetics*, 82(3):247–269, 2000.

D. W. Wohlers and F. Huber. Processing of sound signals by six types of neurons in the prothoracic ganglion of the cricket *Gryllus campestris L.* *Journal of Comparative Physiology*, 146:161–173, 1981.

Combining several sensorimotor systems: from insects to robot implementations

Barbara Webb
Institute for Perception, Action and Behaviour
School of Informatics, University of Edinburgh

Abstract Animals and robots need to integrate different sensorimotor systems to behave successfully. We added an optomotor system to our sound-localising robot, and investigated algorithms for combining the two behaviours. It has been claimed that crickets simply 'add' the outputs of the two responses. We show that several other explanations equally account for the cricket data, and that inhibitory interactions between the behaviours are successful for robot control.

1 Introduction

A general problem in understanding the mechanisms underlying animal behaviour is the integration or interaction of different sensorimotor systems. In the previous chapter we considered the robot implementation of the sound localising behaviour of the cricket. This could be reproduced with a small network of neurons. Crickets have many more neurons (around 100,000) but of course they perform many additional behaviours. Even during phonotaxis, other sensorimotor systems are still active. Böhm et al. (1991) investigated the interaction between phonotaxis behaviour and the cricket's response to visual stimuli. For example, they recorded the angular velocity produced by a cricket walking on a treadmill when played sound from one direction while experiencing a rotating visual field. The results of their experiments led them to conclude that the cricket shows a "turning tendency that can be explained as the weighted sum of the turning tendencies evoked by the two individual stimuli" (see figure 1).

A rotating visual field is, under natural conditions, usually a signal that the animal is itself rotating. By turning in the direction of visual rotation it can correct for whatever forces are causing the rotation. This is known as the 'optomotor' response and is common to many animals. The cricket might use the optomotor response to help stabilise its path when approaching a sound. For example this may explain the faster and more accurate tracking seen under lit vs. dark conditions (Weber et al., 1981). By adding an optomotor sensor to a robot that performs phonotaxis we could explore the interaction of these two sensorimotor systems.

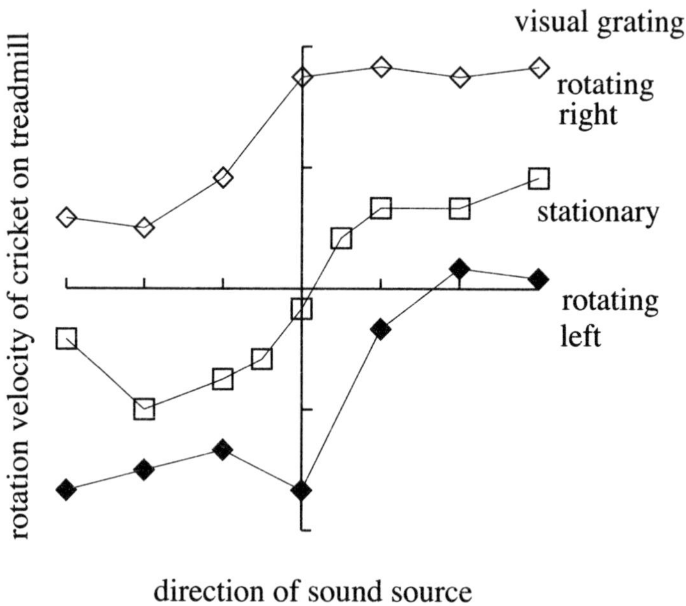

Figure 1. Rotation response of a cricket to combinations of auditory and visual stimuli. When the visual surround is stationary, the walking direction varies sinusoidally with the sound direction. Rotation of the visual surround shifts this curve in the direction of the optomotor response. Based on Böhm et al. (1991)

2 Adding an optomotor sensor to the cricket robot

Harrison and Koch (1998) built a robot that could reproduce insect optomotor behaviour using an analog circuit to perform efficient visual processing. The circuit has a 24 x 6 array of photoreceptors. A local measure of motion is computed between adjacent pairs of photoreceptors in each of six rows across the array as follows. First, the signals are bandpass filtered to remove the DC illumination levels. Each photoreceptor signal is then delayed, using the phase lag inherent in a lowpass filter, then correlated with non-delayed signals from neighboring photoreceptors using multiplier circuits. This constitutes the Hassenstein-Reichardt model of motion detection (Hassenstein and Reichardt, 1956). The results are summed, and the output lowpass filtered to remove residual pattern dependencies from the response. All of these operations are performed on a single analog VLSI chip that dissipates less than 1 mW of power. The output is a voltage signal that increases for rightwards motion and decreases for leftwards motion.

It seemed straightforward to combine this sensory system onto the robot platform we had used to reproduce phonotaxis (see previous chapter) and use a weighted sum of the two sensory system outputs to control its behaviour. However, we soon found that the optomotor response appeared to be interfering with the phonotaxis behaviour. Essentially the problem was that each turn towards the sound would produce a clear optomotor stimulus, which would cause the robot to 'correct' itself and turn away from the sound again. This unsatisfactory result was an empirical demonstration of the problem theoretically formulated by von Holst and Mittelstadt 1950: how can an animal with an optomotor reflex make intentional turns without automatically correcting (and thus negating) them? This problem was not encountered by the crickets in the Böhm et al. (1991) study because their behaviour was measured under open-loop conditions, which would not produce this environmental feedback.

One obvious and easily implemented solution to this problem is to have the turning response to sound temporarily suppress the optomotor response. This kind of 'switching' behaviour has been shown in several animal systems e.g. in response to escape signals in the locust and during pursuit turns in the housefly. Can it be used to explain the cricket's behaviour?

3 Combining behaviours using inhibition

We first implemented this solution using the following simple algorithm (this replaced the neural circuit for motor control described in the previous chapter). Each motor had a default forward speed (set to 20 in the following experiments). A variable 'ears-signal' was set to +1 when a right turn was signalled by the output of the phonotaxis circuit and -1 for a left turn. This was multiplied by a gain of 20, added to the left motor speed and subtracted from the right motor speed. Thus a right turn signal would make the right wheel stop and the left rotate at double speed; and vice-versa for a left turn. If the ears-signal was 0, i.e. no phonotaxis turn was taking place, the motor speeds were instead modulated by the output of the optomotor circuit. Rightwards visual motion would cause a proportional increase in the speed of the left motor and decrease in the right motor; and vice-versa for leftward motion. The gain for this optomotor response was tuned to produce good visual stabilisation under normal conditions. If the ears circuit was signalling a turn, the optomotor gain was set to zero, i.e. this behaviour was inhibited.

We carried out a series of trials to compare the phonotaxis behaviour with and without the optomotor response (the methods are described in detail in Webb and Harrison (2000)). Under normal conditions, the optomotor response appeared to make the tracks somewhat more direct to the sound (figure 2) although this did not turn out to be significant when analysed using a 'directness' measure derived from cricket experiments (Schul, 1998). We did not get the large reduction in range of heading angles reported by Weber et al. (1981) for crickets walking in the light vs. the dark. This might indicate that the cricket is also doing visual tracking, not just optomotor correction, in this situation.

Another reason may be that the robot has rather reliable motor control: if instructed to move in a straight line it will do this fairly accurately without any sensory feedback. Most crickets walking on a treadmill show some systematic directional bias ; and crickets in their natural environment often have motor asymmetries, as well as having to deal with environmentally caused deviations. Thus we carried out further trials after adding a constant bias to the robot's movement. With no sensor input, this would drive the robot on a curved path to the right of the speaker. Without the optomotor response, the robot would tend to head to the right of the sound source when doing phonotaxis, and would fail to reach it on half the trials. When the optomotor response is added (figure 3) there was a significant improvement in the behavior; the tracks did not differ significantly from those produced when there was no motor bias (figure 4).

4 Alternative solutions?

The inhibition mechanism was simple to implement and has some biological plausibility. Nevertheless there are alternative schemes, which also have some biological support. Collett (1980) describes several: 'efferent copy', in which the expected signal optomotor signal resulting from a turn is subtracted from the actual signal; 'follow-on' in which the intended turn is actually controlled via the optomotor response by injecting the inverse of the expected signal, so that the optomotor system in correcting for the apparent signal executes the desired turn; or simply to make the size of turn larger by an amount that should compensate for the expected optomotor response.

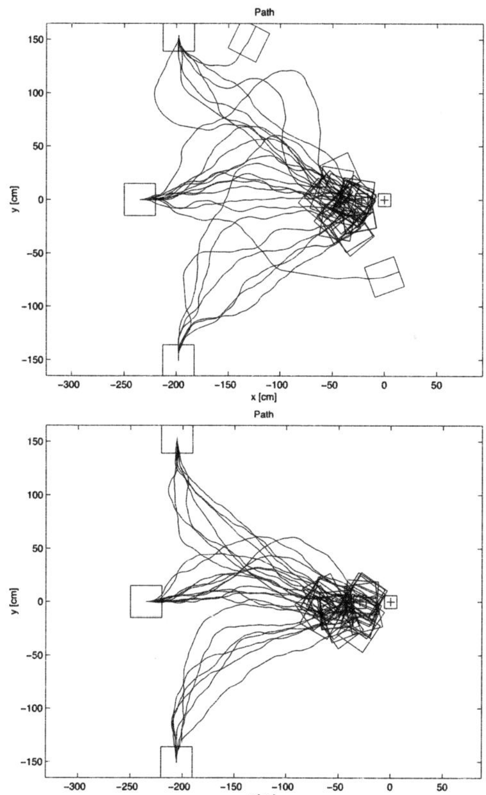

Figure 2. Tracks of the robot to a sound source: left, with no optomotor reflex; right, with the optomotor reflex

As Collett (op cit.) shows, these three schemes algorithmically all reduce to addition of the two signals with appropriate gains. This implies that the simple additive system might have worked if we had simply scaled the gain for phonotaxis to compensate for the optomotor effect. However as Collett also shows, the schemes are not equivalent when considered at the more detailed level of the temporal dynamics of the different reflexes. This was demonstrated by the fact that we could not find suitable additive gain parameters for the robot. For example, turning twice as fast in response to sound simply led to a stronger optomotor signal, which because of the inherent delays in the optomotor low-pass filter, tended to reach its maximum just as the phonotaxis turn ended. In fact, even in the implemented

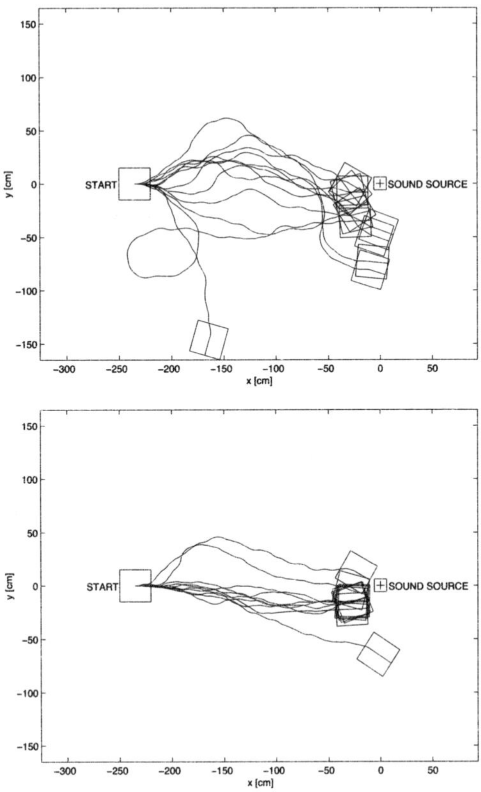

Figure 3. Tracks of the robot to a sound source when it has a motor bias to drive right: left, with no optomotor reflex; right, with the optomotor reflex.

inhibition system, this 'residual' activity in the optomotor signal tended to make the robot turn back a little after each turn. This is because the visual signal was still being integrated during the turn even though the response was inhibited. It would seem a better idea to suppress the signal before the integration stage. We did not originally do this on the robot because the integration was occuring in hardware.

If we add these two possibilities for inhibition - 'pre-integration' and 'post-integration' - to the three alternatives described by Collett (efferent copy, follow-on, scaled addition) we have five possible algorithms for combining the optomotor and phonotaxis responses. We investigated how well each might account for the original cricket results by simulating the open-

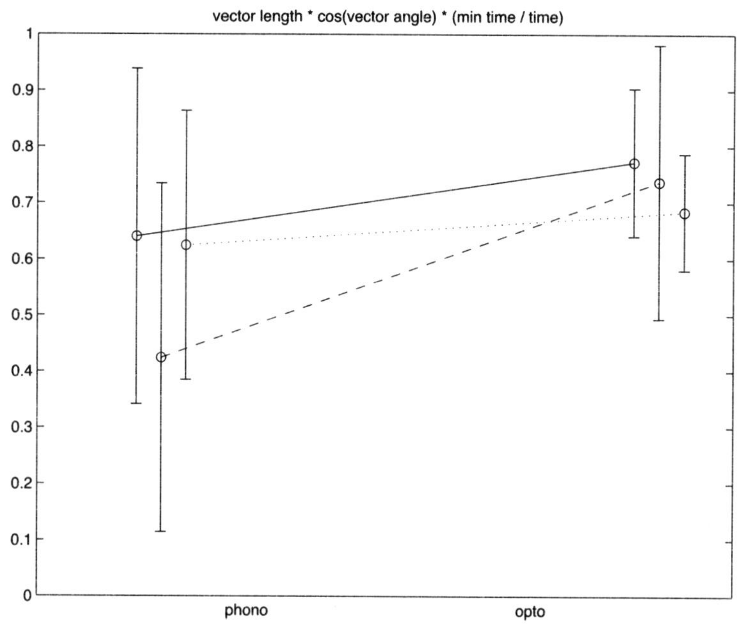

Figure 4. Directness of tracks to a sound source, with and without an optomotor response. Each track is described by a mean vector whose length and angle reflect how directly the track leads to the sound. These are combined with the time taken (as shown in the equation above the plot) to give a 'directness' value for each track (Schul, 1998). Plotted are the mean and standard deviations of directness for each set of ten tracks recorded under different conditions: solid line - trials starting from the centre; dotted line - trials from the sides; dashed line - trials with added motor bias.

loop paradigm that was used by Böhm et al. (1991).

5 Simulating the alternatives in open loop

In the original experiments, the cricket was fixed above an air-suspended ball, so that any turning movements it made would be recorded as angular velocity of the ball. Forward movement does not bring it any closer to the sound source. For the simulated cricket, at each point in time, the position of the sound source and the optomotor stimulation are used to calculate the angular velocity for the next time step. Details of the exact algorithm can be found in Webb and Reeve (2003). The mode of integration of the sound and optomotor signals can be chosen from the following six possibilities:

0. 'no opto' - any optomotor input (imposed or self-generated) is ignored. This corresponds to the pure phonotaxis experiments done on the robot.

1. 'additive' - the two inputs are simply added with appropriate gains. Compared to no opto, the ear gain was doubled to try to compensate for the expected optomotor signal.

2. 'Pre-integration inhibition' - when turning in response to a sound signal, the optomotor signal is inhibited at the input to a low-pass filter.

3. 'Post-integration inhibition' - when turning in response to a sound signal, the optomotor signal was inhibited at the output of a low-pass filter. This corresponds to the integration algorithm previously tested on the robot.

4. 'Efferent copy' - when turning in reponse to a sound signal, a corresponding opposite signal was added to the optomotor signal to cancel out the expected input that would be generated by a turn.

5. 'Follow on' - turning to sound signals was implemented indirectly by adding double the opposite signal to the optomotor signal so that the required turn would occur as part of the optomotor response.

A constant imposed optomotor stimulation could also be added to the opto-input. Also, it was assumed that the phonotaxis response would be intermittent because the normal sound signal is intermittent (i.e. there are gaps between syllables and chirps, and the robot doesn't always detect the sound). This was simulated by having the ear input set to 0 (whatever the direction) for 10 cycles out of every 20 (note 10 cycles in the simulator were roughly equivalent to 100 cycles in the robot, thus one simulated cycle represents about 10ms of real time).

Each integration mode was tested under nine possible combinations of three sound directions (+60, 0 or -60 degrees) and three imposed optomotor stimulus values (+1, 0 and -1). Figure 5 shows the resulting average angular velocity over 200 cycles. Several things are obvious from these plots. First is that the pattern of response under the additive, efferent copy, and follow-on modes is identical; in each case the visual rotation shifts the turning rate in the expected direction. Second, the pattern of the response under the two inhibition modes is identical but not the same as the other schemes. Rather than a uniform increase or decrease in rotation velocity in response to the imposed optomotor signal, the effect 'levels off': that is to say, the mean rotation velocity is the same for 0 or +60 degree sound when there is a positive optomotor stimulus, and for the 0 and -60 degree sound when there is a negative optomotor stimulus. The response is continuous turning at the same rate, either because there is only optomotor driven turning (at 0 degrees) or the phonotactic turns are alternating with the optomotor turns (at 60 degrees). Finally (and not surprisingly) the optomotor rotation has no effect in the 'no-opto' situation.

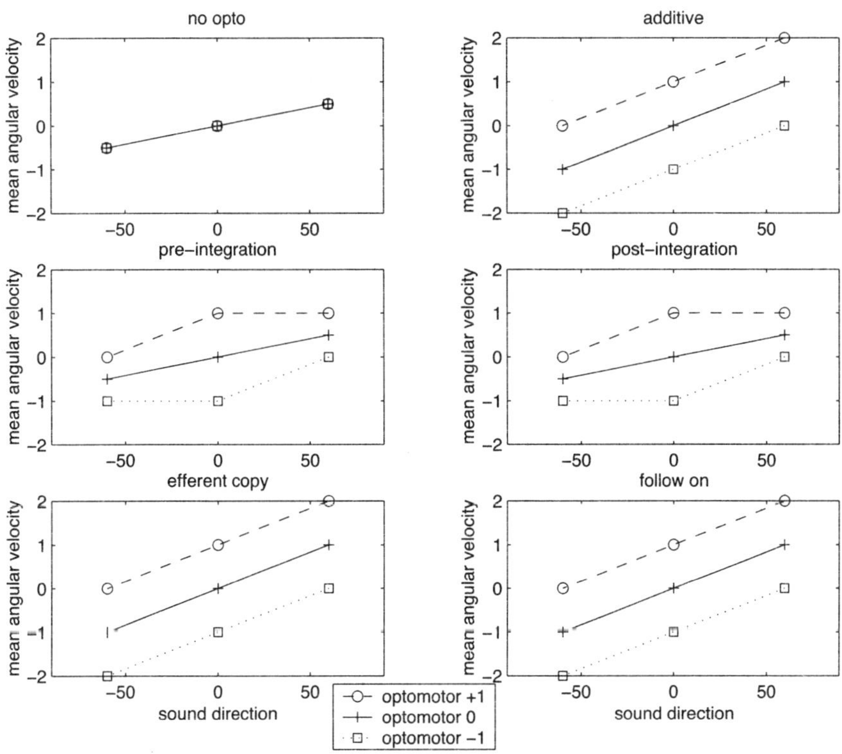

Figure 5. Simulated open loop behaviour under different integration schemes (compare to figure 1). The effects of added visual rotation are identical for the additive, efferent copy, and follow-on algorithms, but the pre-integration and post-integration inhibition mechanisms more closely resemble the results for the cricket.

6 Conclusions

The simulation demonstrated that the open-loop paradigm used in cricket experiments does not distinguish between the additive, efferent copy and follow-on algorithms. The cricket data does not, on close inspection, appear to be a simple addition of a constant, optomotor-induced, shift to the response curve induced by sound as originally claimed by Böhm et al. (1991). The effect looks more similar to that seen in the results of the simulated inhibition schemes. However, it could also be accounted for by any of the other schemes with the addition of an assumption of maximum turning velocity. The resemblance to the inhibition schemes occurs because we assume the animal in normal phonotaxis is regularly switching between sound-following behaviour and optomotor behaviour. However, there is no clear data for the cricket to say how continuous or otherwise is its response to sound, e.g. is it still turning during the gaps between chirps?

These results reveal some of the pitfalls in trying to reason from behavioural results to underlying mechanisms of control. Using time averaging to summarise data, in this case, made it impossible to distinguish whether the insect is switching between behaviours (which are averaged in the results) or actually performing an averaged behaviour in response to two stimuli. Closer analysis of the behaviour (such as looking at directional distributions) should make it possible to distinguish between these possibilities, but may not easily reveal the differences between averaging, efferent copy and follow-on control. The latter might be distinguished by careful considerations of behavioural dynamics - i.e. if the animal can respond much faster to sound than optomotor stimuli, then it seems unlikely the sound response is being controlled via modulation of the optomotor response. It should also be possible to test for efferent copy by investigating the insect's response to incorrect feedback. However such experiments might be complicated by indications that insects can quickly detect when they are in open vs. closed loop and adjust their behaviour accordingly (Heisenberg and Wolf, 1988).

More recently we have looked at using neural implementations of the different algorithms on the robot (Reeve and Webb, 2002). Additive integration was straightforward as the outputs from phonotactic and optomotor interneurons could both modulate the excitation of the same motor neurons to influence turning behaviour. Inhibitory integration could have be implemented in several different ways. The most straightforward was to have the phonotactic output make an additional inhibitory synaptic connection onto the appropriate optomotor interneuron. However, this results in a scheme that falls somewhere between inhibition and efferent copy. Inhibition is represented in our model by a biologically realistic 'shunting' mechanism, which will counteract excitation up to a certain level. Thus, this form of inhibition corresponds to 'ignoring' optomotor excitation up to the amount expected during a turn, in the direction expected during a turn. It thus begins to resemble the efferent copy scheme, although it lacks the subtlety of attempting to exactly predict the size and time-course of optomotor excitation, which a proper efferent copy mechanism would need to do. The follow-on scheme could be implemented in a fairly direct way by having the phonotactic output directly excite the optomotor interneuron that would lead to the appropriate turn.

This latter scheme proved problematic when tested on the robot simply because the output signals from the phonotactic circuit were not appropriately scaled for the optomotor system, so it responded more rarely to sound. While this could have been adjusted in our model, it points out the inherent unlikeliness of this solution as a practical hypothesis for the animal: to have its responses to one modality filtered by the parameters and time-constants of another modality does not seem an effective mechanism. It seems particularly unlikely in this case, where crickets are very capable of performing the sound localisation behaviour in the dark. Of the remaining alternatives, the additive scheme revealed the same problem as in original trials i.e. the optomotor response tended to fight against the intended turns of the robot, resulting in curved tracks from the sides, and worse behaviour when random turns were introduced. By contrast, the inhibitory/efferent copy scheme was very effective in increasing the directness of the robot's tracks under both normal and randomly-disturbed conditions.

We would therefore suggest that some variant of inhibitory interaction is currently the most plausible hypothesis for how the cricket integrates phonotaxis and optomotor behaviour. This leaves us with a number of intriguing questions. Is it possible that insect brains actually carry out 'forward-modelling' to produce efferent-copy cancellation of expected input? Many current theories of primate motor control use such models, but as yet it is not clear if the same arguments really apply to insect behaviour. What neural circuitry might support such functions? These issues will form the focus of our future work.

Bibliography

H. Böhm, K Schildberger, and F Huber. Visual and acoustic course control in the cricket *Gryllus-bimaculatus*. *Journal of Experimental Biology*, 159: 235–248, 1991.

T. Collett. Angular tracking and the optomotor response: an analysis of visual reflex interaction in a hoverfly. *Journal of Comparative Physiology*, 140:145–158, 1980.

R. R. Harrison and C. Koch. An analog VLSI model of the fly elementary motion detector. In M.I. Jordan, M.J. Kearns, and S.A. Solla, editors, *Advances in Neural Information Processing Systems 10*, pages 880–886. MIT Press, Cambridge, MA, 1998.

B. Hassenstein and W. Reichardt. Systemtheoretische Analyse der Zeit-

, Reihenfolgen-, und Vorzeichenauswertung bei der Bewungsperzeption des Rüsselkäfers. *Chlorophanus. Z. Naturforschung*, 11b:513–524, 1956.

M. Heisenberg and R. Wolf. Reafferent control of optomotor yaw torque in *Drosophila-melanogaster*. *Journal of Comparative Physiology A*, 163(3): 373–388, 1988.

R. Reeve and B. Webb. New neural circuits for robot phonotaxis. In *EPSRC/BBSRC International Workshop Biologically-Inspired Robotics: The Legacy of W. Grey Walter*. Hewlett-Packard, 2002.

J. Schul. Song recognition by temporal cues in a group of closely related bushcricket species (genus tettigonia). *Journal of Comparative Physiology A*, 183:401–410, 1998.

B. Webb and R.R. Harrison. Integrating sensorimotor systems in a robot model of cricket behavior. In *Sensor Fusion and Decentralised Control in Robotic Systems III*. SPIE, Boston Nov 6-8, 2000.

B. Webb and R. Reeve. Reafferent or redundant: How should a robot cricket use an optomotor reflex? *Adaptive Behaviour*, 11:137–158, 2003.

T. Weber, J. Thorson, and F. Huber. Auditory behaviour of the cricket I Dynamics of compensated walking and discrimination paradigms on the Kramer treadmill. *Journal of Comparative Physiology A*, 141:215–232, 1981.

Sensory Feedback in locomotion control

Paolo Arena, Luigi Fortuna, Mattia Frasca and Luca Patané
Dipartimento di Ingegneria Elettrica Elettronica e dei Sistemi,
Universitá degli Studi di Catania, Catania, Italy

Abstract This chapter focuses on the sensing processes and their interactions with locomotion control. The analysis has been accomplished taking into account different levels of behavior. Moreover dynamic simulators and robotic structures have been used to investigate the biological principles governing the sensory feedback in the real world.

1 Introduction

All living beings, bacteria, plants, animals possess sensors and sensing processes. Animals live and move in a world where the environment is constantly changing. They have an abundance of sensors to monitor both body movements and the surroundings in order to adapt the behavior according to the sensorial stimuli.

Animal behavior can be roughly categorized into three mayor classes (Arkin, 1998):

Reflexes - They are rapid, automatic involuntary responses triggered by external stimuli. The response persists for the duration of the stimulus and its intensity is correlated with the stimulus's strength. Reflexes are very useful during locomotion, making the animal able to rapidly react producing, for example, escape behavior such as that found in locust when a collision is expected.

Taxes - They are behavioral responses orienting an animal toward or away from an attractive or aversive stimulus. Taxes are associated to several phenomena such as visual, chemical, mechanical. An example of chemotaxis is the trail following of ants.

Fixed-action patterns - They are time-extended action patterns triggered by a stimulus but persisting for longer than the stimulus itself. The intensity and duration of the response are independent by the stimulus. In contrast with reflexes, fixed-action patterns may be governed not only by environmental stimuli but also by the internal state of the animal (e.g. influenced by appetite). Examples include the courtship ritual that precedes mating in numerous species of insects.

These different behavior mechanisms permit a continuous interactions between the living organisms and the environment in order to improve their locomotion capabilities.

Another important characteristic is an enormous redundancy of sense organs. In consequence, the loss of any single sense organ rarely has even minor effects on walking (Cruse et al., 1984). This may aid adaptability as well as allow an insect to compensate for damage to its legs.

Taking inspiration from the animal world, several capabilities could be included in a robot in order to adapt it to the real world in the same way. The integration of sensory feedback on robots may introduce a substantial improvement in different application fields: mechanical preflexes and sensory reflexes may help stabilize walking (Dickinson et al., 2000; Delcomyn, 1991), environment feedback is fundamental when a robot is engaged in climbing obstacles (Watson et al., 2002a,b) and taxis is important for the navigation control in order to realize appetitive or aversive behaviors (Webb and Consi, 2001).

During the analysis of different sensors processes, taking inspiration by nature, special attention has been devoted to insects, which are able to show several motor reflexes. In fact, one of the main characteristics of arthropods such as insects, spiders and scorpions is their cuticula exoskeleton. It plays an important sensory role: hair-like sensors are distributed in the whole body and recognize air flow or touch, furthermore the detection of cuticular strains is accomplished by the campaniform sensilla (i.e. tiny holes able to characterize the acting force (Humphrey et al., 2003)). Tactile hair sensors are adopted for exteroceptors stimuli from outside the animal and also proprioceptor which sense stimuli produced during locomotion by the animal itself like the contact between neighboring body parts (e.g. sense the joint angle between the adjacent segment of a leg). The information produced by this network of sensors is devoted to actuate adaptive mechanisms.

To adapt in difficult environment, the robot should have artificial senses so that it can acquire information and react to the sensing. Taking into account the arthropods, different sensing capabilities can be identified (Dickinson et al., 2000; Kingsley et al., 2003).

In the following sections different classes of sensorial stimuli will be discussed, focusing on the feedback with the low level system for the locomotion control in order to obtain a response.

2 Sensory feedback in locomotion control

Regarding the navigation of walking robots, a basilar problem is the control of movement direction. The trajectory that a system would follow can be planned at high level and then adapted at low level to execute obstacle avoidance and target detection procedures. Taking inspiration from the animal world, different levels of behavior can be distinguished. Three experiments, related to the three behavior classes (Arkin, 1998) previously introduced, will be proposed in order to underline the sensing capabilities of a bio-inspired robot.

A low level control corresponding to a reflex behavior is used by the robot to face with the obstacle avoidance problem. In the second experiment a taxis mechanism is adopted to make the robot able to follow a target. In the last experiment an emerging behavior, observed during a target approaching action, is reported. The robot response can be classified as a fixed-action pattern.

Moreover the integration between sensory feedback and motor system as in Cohen and Boothe (1999), will be described.

2.1 The Control System

The direction control is a fundamental aspect of mobile robotics. An important application is obstacle avoidance. The analysis of the problem could be divided into two parts: the introduction of a sensorial system able to identify an obstacle and the coupling strategies between the sensorial response and the locomotion system.

The approach here proposed has been inspired by the well-known Braitenberg vehicles (Braitenberg), where the sensors are directly coupled to the motors. In our case the coupling is realized with connections between sensors and leg controllers (i.e. CPG cells) that are also connected each other. The Braitenberg vehicles are simple dual drive robots able to reproduce an attractive or repulsive behavior. The speed of each wheel is influenced by the sensory responses that can be increased or decreased producing a steering behavior. This control strategy has been applied to a six legged-robot (Arena et al., 2002b). The sensory stimulus inhibits one of the middle legs of the hexapod robot; the leg stopped in the stance phase is then used as pivot to change the movement direction.

2.2 Experiment 1: Obstacle Avoidance

The obstacle avoidance strategy was firstly validated taking into account only the control system. The suitability of the approach is guaranteed by

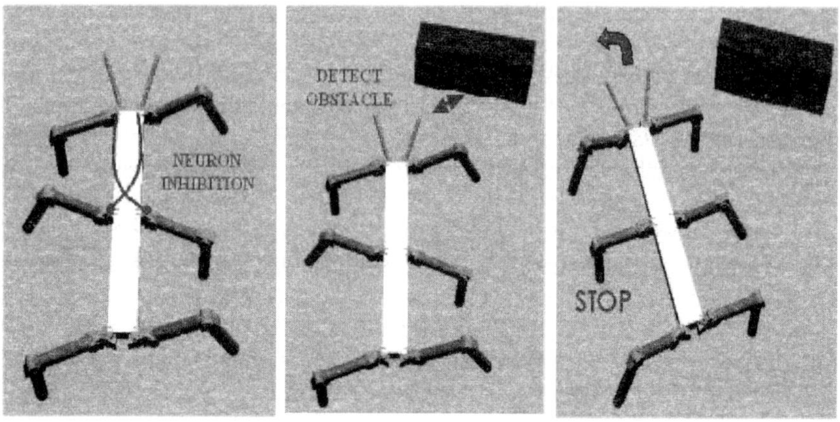

Figure 1. Behavior of a hexapod model, realized in Visual Nastran 4D, when the obstacle avoidance procedure is active.

the fact that the sensor feedback modifies the locomotion pattern without dramatic changes in the behavior of the cells not directly coupled to the sensors.

Moreover a dynamic model of an hexapod robot, realized in Visual Nastran 4D, was adopted. The obtained results show the capability of the system, able to avoid an obstacle along the movement direction. The robot antennae identify the object when the distance is under a given threshold. The fundamental aspects of the obstacle avoidance procedure are shown in Figure 1.

In the last step the sensory feedback was tested on an hexapod robot prototype, named *PChex*, where the locomotion pattern is given by a PC.

The direction control, implemented on the robot, is based on the inhibition of oscillations of the CNN neurons controlling the middle legs as discussed in Arena et al. (2002b). The two antennae, as in real insects (Durr and Krause, 2001), detecting the presence of obstacles to be avoided are implemented each one by a pair of contact switches. The switches are connected by a thin rubber pipe and the sensor output is the logical OR of the two switch outputs. This allows to detect obstacles in a 180° orientation range. When one of the sensors is activated, the corresponding leg is stopped for a fixed short time interval, allowing the robot to avoid the obstacle. Figure 2 shows the trajectory obtained while the robot is walking on a corridor (the wall on the right side and the wooden panels constitute

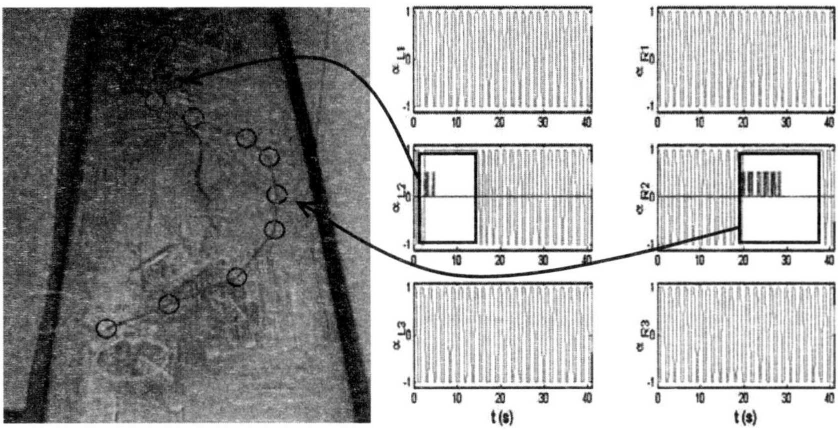

Figure 2. Direction control of the robot: trajectory of the robot and control signals for the swing motor of each leg (for legs R2 and L2 the sensor outputs are also shown in the panel).

the obstacles).

The robot is able to identify obstacles with a pair of infrared sensors. The system is also equipped with two antennae following the redundancy strategy typical of animals. When the distance sensors have a fault, the mechanical switches can identify the obstacle and allow collision avoidance.

The introduction of a reflexive behavior makes a legged-robot able to walk in an unstructured environment avoiding collisions.

2.3 Experiment 2: Target Following

Adopting the Braitenberg's approach (Braitenberg), a taxis behavior has been reproduced in a bio-inspired robot in order to follow a moving target. The attractive behavior was validated by using the first version of the *Hexchip* robot. To make the robot able to reach a given target or to follow it, a sensor system has been designed. The target, shown in Figure 3 (a), is an infrared emitter diode modulated at a frequency of $2kHz$. The robot has been equipped with three receivers, one is shown in Figure 3 (b): they are constituted by a photo-diode and a PLL (LM567) locked to the emission frequency. The sensors are located in the two sides and in the back of the hexapod robot.

When the target is within the sensor action range, the direction control

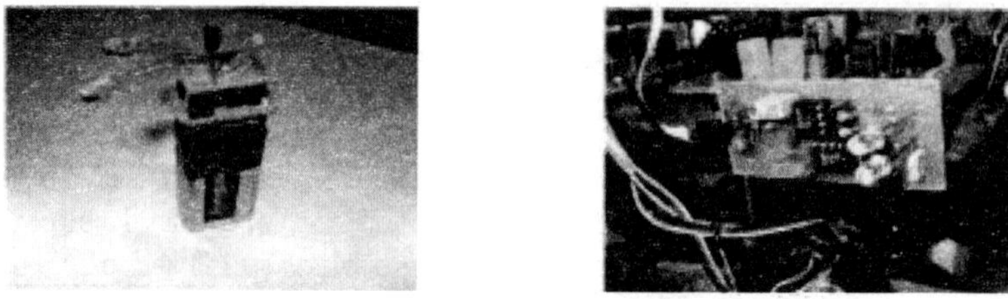

(a) (b)

Figure 3. (a) A modulated infrared emitter diode realizes the target. (b) The receiver consists of a photo-diode and a PLL locked to the emission frequency.

procedure modifies the robot locomotion; when the sensors perceive no signals, the target is in front of the robot that proceeds in forward direction. During the experiments, the *Hexchip* robot was able to follow a moving target as shown in Figure 4.

2.4 Experiment 3: Emerging Behavior

The last experiment outlines an emerging behavior observed in the *Hexchip* robot during a target detection procedure. The target is fixed on the floor and the robot is deployed in an arbitrary position, a few meters far from it. The robot attracted by the target, proceeds towards it, but, when the distance is small enough, the target produces an omnidirectional attraction. Under this stimulus, the robot moves around the target reproducing a circular path as shown in Figure 5.

The movements of the robot resembled the dance pattern shown by numerous species of insects (Davey; Thornhill and Alcock, 1983) during a courtship ritual that precedes mating. The robot behavior is clearly uncorrelated with these fixed-action patterns, but following the basic approach of Braitenberg we can observe how rich behaviors arise even in simple experiments.

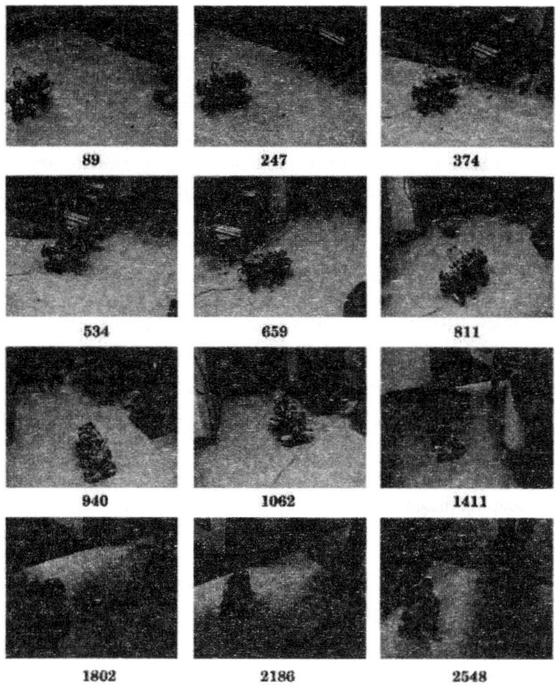

Figure 4. The robot follows a person holding the infrared target. Each snapshot is labelled with the frame number.

Figure 5. Fixed-action pattern, the robot finds the target and moves in circle around it.

3 Sensory Reflexes

As shown in Cohen et al. (1982) the CPG receives signals from higher level neurons that are necessary to initialize the locomotion, but not to generate the pattern of right movements of the effector organs. Analogously the presence of feedback from environment is not strictly necessary to generate a rhythmic pattern; in fact this one has been observed even when the feedback has been inhibited. Obviously, feedback signals modify the locomotion pattern and are fundamental to achieve good performance of the whole control system on real environments. From the observation of the animal behavior it is evident how the interaction between environment and locomotion acts at different levels. The stimuli coming from the environment can influence the locomotion pattern in different ways: a difficult situation can produce a change of the locomotion gait (for example when the insect perceives a danger it uses the fast gait to run away) otherwise a local alteration of the normal locomotion can be solved simply by using local recovery procedures named reflexes. They are locally generated from the nervous endings of a leg and produce an action that modulates the CPG signals.

Recent studies of neurophysiologists identified in insects, three different typologies of reflex (Beer et al., 1997) :

- *Stepping reflex* allows a leg to find a stable support point after sliding.
- *Elevator reflex* allows a leg to get over an obstacle after a hit, increasing the leg elevation.
- *Searching reflex* allows a leg to find a new support point when in the desiderate anterior extreme position (AEP) there is a hole.

Thanks to the local reflexes, animals are able to move on uneven terrain, otherwise unaccessible without an additional overhead on the CPG level or on the high-level control.

The properties of the new control system have been investigated. The principal objective is to underline the interaction between distributed control systems, like local reflexes, and high level control strategies as posture control. At this aim a hexapod robot, designed in a dynamic simulation tool, has been used.

3.1 Elevator reflex in the CNN-based CPG

The role of local feedback in the locomotion of legged robots has been investigated.

When insects walk on irregular terrain, they move in a very effective way adopting a variety of local leg reflexes (Pearson and Franklin., 1984). Insects are able to deal with irregular surfaces, small elevated steps, ditches, poor supports and so on. These reflexes can provide robotic solutions to build up autonomous structures able to negotiate uneven terrain (Klaassen et al., 2002). Indeed, reaching areas that cannot be accessed by wheeled vehicles is a clear advantage of walking robots: this can be achieved by implementing local reflexes.

The idea focuses on the application of elevator reflex in a hexapod model in which the locomotion control is performed by a CPG based on CNNs.

By using the dynamic properties of the motor-neuron, it has been possible to reproduce the dynamics that characterize the elevator reflex in insects.

The advantage of a CPG, realized with a dynamical system, is the capability to modify the system parameters in order to obtain the wished behavior.

The role of feedback in a CPG, devoted to control the locomotion, is essential to deal with complex environments. The elevator reflex, used by insects to overcome a little obstacle after a collision with a leg will be discussed in detail. Moreover the reaction of the whole control system to the introduction of the local feedback has been examined.

Starting from the standard cell equation for CNN-based CPGs, a new input variable has been introduced. This control signal, triggered by a hit, must be able to change the system behavior avoiding the obstacle. The new motor neuron is the following:

$$\begin{cases} \dot{x}_1 = -x_1 + (1+\mu)y_1 - s\ y_2 + i_1 \\ \dot{x}_2 = -x_2 + (1+\mu)y_2 + s\ y_1 + i_2 + \xi(t) \end{cases} \tag{1}$$

where $\xi(t)$ is:

$$\xi(t) = \begin{cases} A_\xi & t_{hit} < t < t_{hit} + \mathcal{T} \\ 0 & \text{elsewhere} \end{cases} \tag{2}$$

ξ is a short pulse triggered by a hit with an obstacle. When such a signal is applied, the behavior of the standard system changes as shown in Figure 6.

The modification of the bias current i_2 involves an alteration of the neuron dynamic from limit cycle to a stable equilibrium point. During an elevator reflex procedure, a leg that hits an obstacle, rapidly retracts to disengage from it, then lifts the foot and swings, increasing the previous elevation, over the obstacle. A study of the system behavior with different amplitude and duration of the current impulse has shown that a value of $A_\xi = -2.2$ and $\mathcal{T} = T/5$, where T is the limit cycle period, are the best solution and in fact, by using this settings, the maximum value of $x_1 = 2.8$ is reached. It is important to notice that the introduction of the reflex produces a delay in the leg movement cycle.

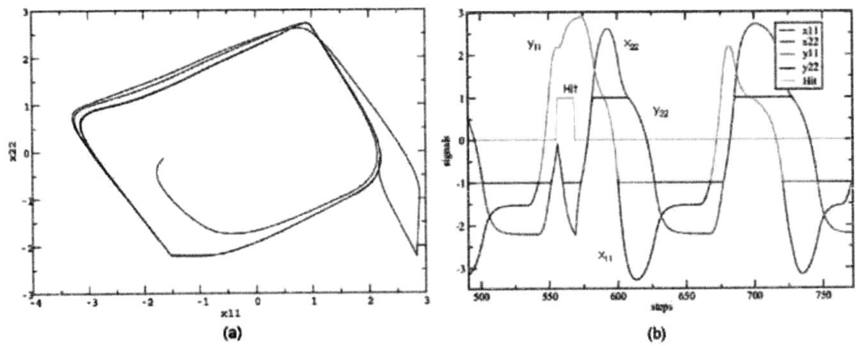

Figure 6. Behavior of a CNN motor-neuron when the elevator reflex is triggered: (a) limit cycle behavior in the phase plane $x_1 - x_2$; (b) time trends of the state and output variables of the motor neuron; the hit signal ξ is also shown.

The variables y_1 and y_2 of each motor neuron were previously chosen to control the elevator/flexor and the protractor/retractor joints of the hexapod robot. To take advantage of the elevator reflex controlling the leg elevation, instead of y_1 another function of the state has been used:

$$z_1 = \begin{cases} x_1 & x_1 > -1 \\ -1 & x_1 \leq -1 \end{cases} \tag{3}$$

The new control variable z_1 is not limited in the upper part, this makes possible to increase the maximum amplitude of the elevation control signal without alteration of the leg stance phase.

The analysis of the effect, due to the introduction of the bias current impulse, previously discussed for a single neuron, is also extended to the whole CPG. The reflex application introduces a delay, and it is important to understand how this disturbance modifies the synchronization in the CPG network. To guarantee the gait stability the pattern synchronization must be maintained or otherwise recovered after a short transient. The delay introduced by the reflex action in the locomotion gait depends on the type of couplings between the neurons and then it is characteristic of each gait.

In the obstacle overcoming problem, the gait adopted by the robot may be able to face with difficult terrains and then it must be characterized by a high stability margin. The formula, proposed in Ting et al. (1994), to compute the longitudinal stability margin (S) of a wave gait (i.e. a regular, symmetrical gait) for a legged robot is reported below:

Table 1. Stability margin for the three patterns obtained with the CNN-based CPG

Gait	Slow	Medium	Fast
Duty Factor	$\frac{9}{12}$	$\frac{5}{8}$	$\frac{1}{2}$
Stability Margin	0.375	0.1875	0

$$S = \frac{n_{legs}}{4}\beta - \frac{3}{4} \tag{4}$$

where β is the duty factor, the fraction of a stride period that a leg is in the support phase and n_{legs} represents the number of legs. When $S < 0$, the robot is statically unstable.

The stability margins obtained applying this criterion to the fast medium and slow gaits typical of the CNN-based CPG are given in Table 1.

In the following examples the Medium gait is adopted, it represents a good balance between locomotion stability and speed. In fact because each time four legs contact the ground, when a leg is involved in the reflex procedure a minimum delay due to the elevator reflex resolution is not critical for stability.

3.2 Simulation Results

The validation of the elevator reflex control scheme has been accomplished on a dynamic hexapod robot model implemented by using DynaMechs simulation libraries.

A set of contact sensors have been distributed on the upper part of the toe of each leg. The hit signal for each leg is obtained with a logic OR between its sensors. To avoid the uncorrect generation of hit signals, due to the terrain roughness, the sensors are activated only during the early swing and the swing phase and are deactivated during the late swing and the stance phase.

A very efficient way to realize a bump sensor, able to identify a collision with an obstacle in a hexapod robot, is reported in Klaassen et al. (2002); Arena et al. (2003b). The main idea is to realize a virtual sensor monitoring the motor status. By measuring the current absorbed by the protractor/retractor joint motor, it is possible to identify whether a leg encounters an obstacle during the swing phase.

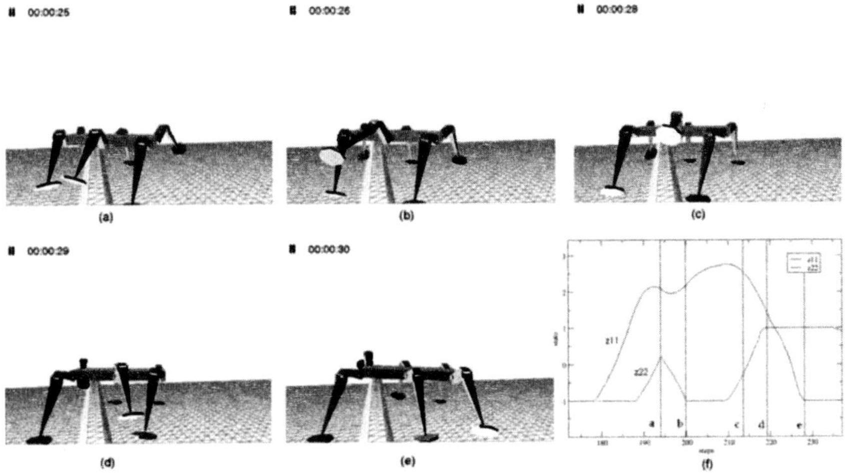

Figure 7. The leg $R2$ of the hexapod model is able to overcome an obstacle thanks to the introduction of elevator reflex. The whole sequence is shown in five snapshots. In the last image on the right the two variables devoted to control the elevator/flexor joint ($z11$) and the protractor/retractor joint ($z22$) are reported. The intervals a, b, c, d and e correspond to the sequence of images.

Obstacle overcoming The final aim of the reflex implementation on the robot is the improvement of its capabilities. In the case of elevator reflex the new ability is the overcoming of small obstacles during the locomotion. The sequence of snapshots in Figure 7 shows the behavior of the robot when the leg $R2$ collides with an obstacle. The height of the obstacles used during the testing is around the $80-90\%$ of the robot height from the ground. The sequence of actions is also described by using the leg $R2$ control signals. The five steps identified are explained in following:

(a) The leg $R2$ hits an obstacle. The bias current impulse triggered activates the elevator reflex.

(b) The leg rapidly retracts to disengage the obstacle.

(c) The leg lifts the foot increasing the previous elevation.

(d) The leg swings over the obstacle.

(e) The leg touches the ground.

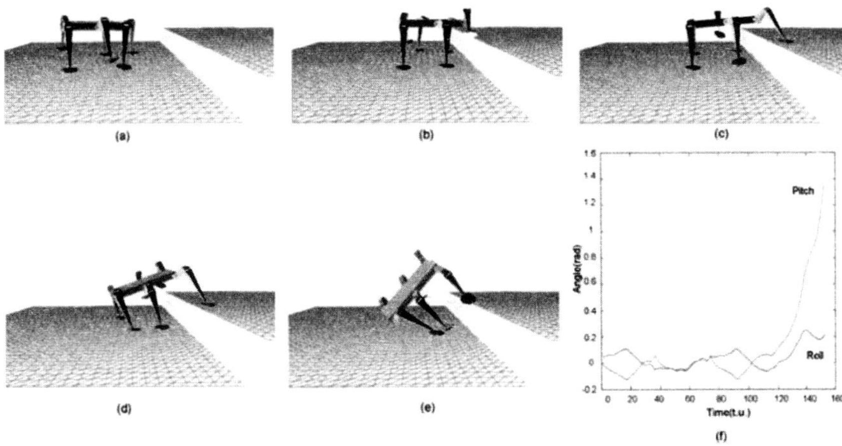

Figure 8. When the hexapod deals with a high step, the low level control (elevator reflex) is not sufficient: the robot is not able to overcome the obstacle and crashes into the ground. In the last image the trend of pitch and roll angles during the simulation is given.

Difficult terrain When the upper part of the obstacle is enough large, at the end of the elevator reflex procedure, the leg could be positioned on the obstacle. To valuate the system behavior in this situation, the robot locomotion on a step has been analyzed. The first trials have been shown that when the anterior legs are positioned on the step, the robot is unbalanced and crashes on the ground, as illustrated in Figure 8.

A low level control, like a local feedback, is not able to deal with this scenario. To solve this problem the introduction of a higher level control has been taken into account. Therefore an attitude control has been integrated in the control system. This control action, that involves the whole system, makes the robot able to move also on uneven terrain. In Figure 9 it is shown how the introduction of the posture control (Arena et al., 2002a, 2003a) makes the robot able to maintain its body parallel to the horizontal line. Therefore the stability of the robot, applying either the high and low level control, is guaranteed also in overcoming a high step. The principle of integration between the Central Nervous System (CNS) and the low

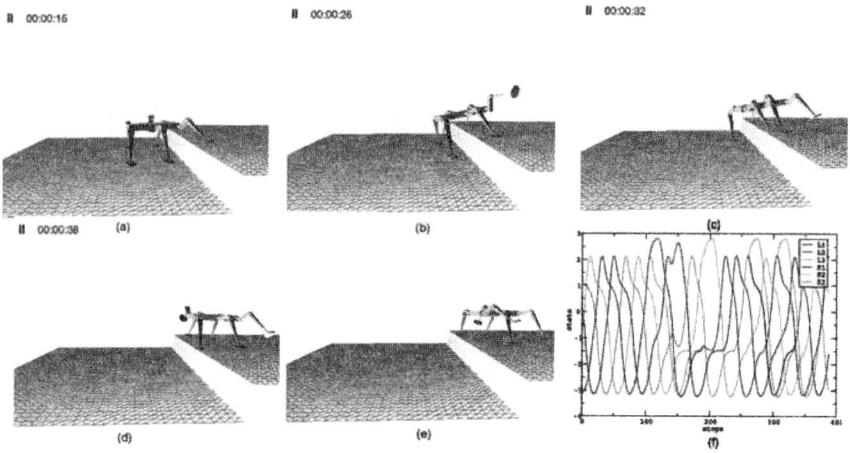

Figure 9. When the hexapod deals with a step, the integration of a high level control (attitude control) and a low level control (elevator reflex) produces a synergy that makes the robot able to overcome the obstacle. In the last image the variables x_1 of each leg show the sequence of reflexes activated by the hit with the obstacle.

level feedback, as reflexes, was demonstrated in several experiments with animals (Prochazka et al., 2002) and used as a paradigm in the design of control system for bio-mimetic robots (Quinn and Ritzmann, 1998; Nelson and Quinn, 1999). In the example of obstacle climbing, the application of an attitude control improves the system stability during the motion phase allowing to adapt the robot movement to the terrain morphology. The dynamic simulations carried out show that when the hexapod walks on difficult terrains, the local feedback is not enough and a higher level control must be also applied. A strictly collaboration between the CNS and the low level control is the basis of emerging behaviors that arise when a difficult situation can not be solved with a low level feedback.

Bibliography

P. Arena, L. Fortuna, and M. Frasca. Attitude control in walking hexapod robots: an analogic spatio-temporal approach. *Int. J. Circ. Theor. Appl.*, 30:349–362, 2002a.

P. Arena, L. Fortuna, M. Frasca, and L. Patané. CNN based central pattern generators with sensory feedback. In *7th International Workshop on Cellular Neural Networks and Their Applications*, pages 275–282, Frankfurt (Germany), 2002b.

P. Arena, F. Conti, L. Fortuna, M. Frasca, and L. Patané. Nonlinear networks to control hexapod walking. In *NDES*, 2003a.

P. Arena, L. Fortuna, M. Frasca, L. Patané, S. Testa, and L. Zagarella. A tele-operated walking hexapod controlled by a CNN-based CPG. In *Int. Conf. on Climbing and Walking Robots*, September 2003b.

R. C. Arkin. *Behavior-Based Robotics.* MIT Press: Cambridge, MA, 1998.

R.D. Beer, R.D. Quinn, H.J. Chiel, and R.E. Ritzmann. Biologically inspired approaches to robotics. *Communications of the ACM*, 40(3), March 1997.

V. Braitenberg. *Vehicles: experiments in synthetic psycology.* MIT Press: Cambridge, MA, 1984.

A. H. Cohen and D. L. Boothe. Sensorimotor interactions during locomotion: principles derived from biological systems. *Autonomous Robots*, 7: 239–245, 1999.

A. H. Cohen, P. J. Holmes, and R. H. Rand. The nature of the coupling between segmental oscillators of the maprey spinal generator for locomotion: a mathematical model. *J. Math. Biol.*, 3:345–369, 1982.

H. Cruse, J. Dean, and M. Suilmann. The contributions of diverse sense organs to the control of leg movement by a walking insect. *Journal of Comparative Physiology A*, 154:695–705, 1984.

K. G. Davey. *Reproduction in the insects.* Freeman, San Francisco, 1965.

F. Delcomyn. Leg instability after leg amputation during walking in cockroaches. In *Proc. Third IBRO World Congress of Neuroscience*, volume P10.17, page 83, 1991.

M.H. Dickinson, C.T. Farley, R.J. Full, M.A.R. Koehl, R. Kram, and S. Lehman. How animals move: an integrative view. *Science*, 288:100–106, April 2000.

V. Durr and D. Krause. The stick insect antenna as a biological paragon for an actively moved tactile probe for obstacle detection. In *Int. Conf. on Climbing and Walking Robots*, Karlsruhe, September 2001.

J.A.C. Humphrey, F. G. Barth, M. Reed, and A. Spak. *Sensors and sensing in biology and engineering*, chapter The physics of arthropod medium-flow sensitive hairs: biological models for artificial sensors. Barth, Humphrey, Secomb, springerwiennewyork edition, 2003.

D.A. Kingsley, R.D. Quinn, and R.E. Ritzmann. A cockroach inspired robot with artificial muscles. In *International Symposium on Adaptive Motion of Animals and Machines (AMAM)*, Kyoto, Japan, 2003.

B. Klaassen, F. Kirchner, and D. Spenneberg. *Neurotechnology for Biomimetic Robots*, chapter A biologically inspired approach towards robust real world locomotion in an eight legged robot. J. Ayers and J. Davis and A. Rudolph, mit press edition, 2002.

G.M. Nelson and R.D. Quinn. Posture control of a cockroach-like robot. *IEEE Control Systems*, 19:9–14, 1999.

K. G. Pearson and R. Franklin. Characteristics of leg movements and patterns of coordination in locusts walking on rough terrain. *International Journal Robotics Research*, 3(2):101–112, 1984.

A. Prochazka, V. Gritsenko, and S. Yakovenko. *Sensorymotor control*, chapter Sensory control of locomotion: reflexes versus higher-level control. S.G. Gandevia, U. Proske, D.G. Stuart, kluwer academic/plenum publishers edition, 2002.

R.D. Quinn and R.E. Ritzmann. Biologically based distributed control and local reflexes improve rough terrain locomotion in a hexapod robot. *Connection Science*, 10:239–255, 1998.

R. Thornhill and J. Alcock. *The Evolution of Insect Mating Systems.* Harvard University Press, Cambridge, Massachusetts, 1983.

L.H. Ting, R. Blickhan, and R.J. Full. Dynamic and static stability in hexapedal runners. *J. exp. Biol.*, 197:251–269, August 1994.

J.T. Watson, R.E. Ritzmann, S.N. Zill, and A.J. Pollack. Control of obstacle climbing in the cockroach, Blaberus discoidalis. I. Kinematics. *J Comp Physiol A*, 188:39–53, 2002a.

J.T. Watson, R.E. Ritzmann, S.N. Zill, and A.J. Pollack. Control of climbing behavior in the cockroach, Blaberus discoidalis. II. Motor activities associated with joint movement. *J Comp Physiol A*, 188:55–69, 2002b.

B. Webb and T. R. Consi. *Biorobotics.* MIT Press, 2001.

A looming detector for collision avoidance

Paolo Arena and Luca Patané
Dipartimento di Ingegneria Elettrica Elettronica e dei Sistemi,
Universitá degli Studi di Catania, Catania, Italy

Abstract The visual system is one of the most complex sensory architecture used by animals. The incredible technological progresses in the electronic field, nowadays, have not been able to produce devices with performances comparable to the insect visual system in terms of resolution, speed acquisition and processing. In this chapter an algorithm for the collision detection, inspired to the locust visual system will be presented.

1 Introduction

Animal visual systems are incredibly good examples to provide inspiration for the design of sensory systems. Interesting results have been obtained in the realization of retina-like visual sensors characterized by space-variant resolution (Sandini and Metta, 2003) and in the implementation of visually guided robots taking inspiration from the fly vision system reproduced in hardware as a distributed network of elementary motion detectors (Franceschini, 2003).

An important task for moving robots is the detection of looming or motion in depth. Information about approaching objects is highly significant to many animal species. A bio-inspired solution for this task can be found by observing the locust, named *Locusta Migratoria*, that possesses two specific neurons that respond preferentially to movements directed towards the animal.

A swarm of locusts may contain up to 50 million insects per km^2. In this kind of crowd, collisions might be a real problem. An example of this phenomena is shown in Figure 1.

The locust visual sensor consists of a compound eye able to identify change of luminosity. The distributed information, captured from the environment, is then elaborated by a neuron called *Lobula Giant Movement Detector* (LGMD). The LGMD realizes an integration of visual information generating stimuli when a collision is expected. The LGMD neuron is

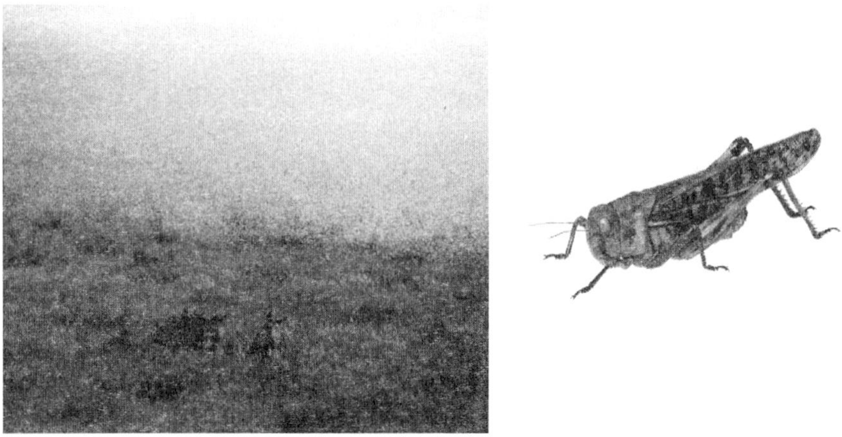

Figure 1. A swarm of *locusta migratoria* in Madagascar, a photo of the insect is shown on the right.

synaptically connected to the *Descending Controlateral Movement Detector* (DCMD) that transfers the LGMD activity to motor centres. Postsynaptic targets for the DCMD neuron include motor-neurons and premotor interneurons that are involved in jumping and flying. Figure 2 depicts the locust visual system.

When an image expands symmetrically over the retina, this is a cue indicating that an object is approaching an animal on a collision course. The LGMD neuron are tightly tuned to detect objects that are approaching on a direct collision course otherwise when a receding object is identified, the avoidance reaction is not triggered.

2 LGMD Neural Network Model

A neural network that respond like the LGMD has been proposed by Rind and Bramwell (1996). The network, schematically illustrated in Figure 3, consists of four layers that represent the organization of the locust LGMD. The first layer contains the P-units, representative of the locust photoreceptors. Each P-unit responds with a brief excitation to a change in the level of illumination marking with great precision the passage of an edge. As in the locust visual system, processing is divided into channels preserving the topological map perceived by the array of photoreceptors. The sensorial stimuli are propagated to the second layer in which can be identified two

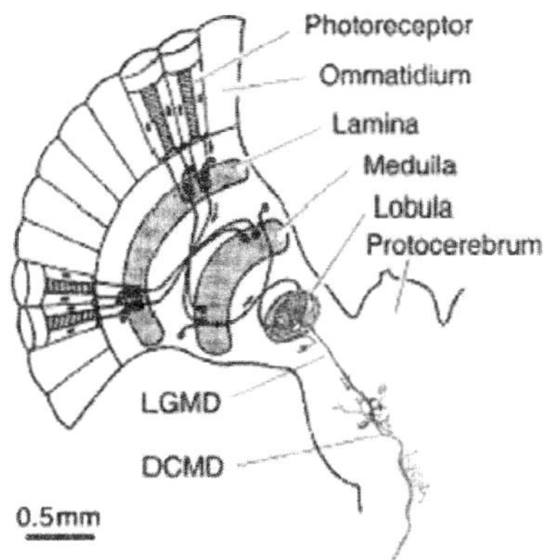

Figure 2. Scheme of the locust visual system. The LGMD and DCMD neurons are responsible of the elaboration of the visual stimulus and the transmission to the motor system, triggering avoidance reactions.

neuronal units: excitatory (E-unit) and inhibitory (I-unit). The E-unit and I-unit follow excitation of the corresponding P-unit producing a spike. A refractory period follows the activation, during which the neuron could not be activated, and excitation decays exponentially. The information processed in layer 2 are transmitted to the summing units (S-unit) in layer 3. Each S-unit collects the response from the corresponding E-unit and from the 6 neighbouring and 12 next-neighbouring I-units with a conduction delay. This inhibition is also weighted with a coefficient equal to 1/6 and 1/12 respectively. The S-unit produces a spike when a given threshold level of excitation is reached in input. Similarly to the other units, immediately following the spike, the S-unit response declines exponentially with time and a refractory period is presented. The final layer of the network (layer 4) consists of a single LGMD-unit that receives and linearly sums all S-unit responses. When a given threshold is reached, an object approaching in a collision trajectory is identified and a motor response is triggered to avoid the imminent collision. A feed-forward inhibitory unit (F-unit) is introduced to distinguish approaching from receding objects. The F-unit is activated when a given number of P-units is simultaneously active. The response of F-unit bypassing layer 3 inhibits after a delay, the LGMD neuron.

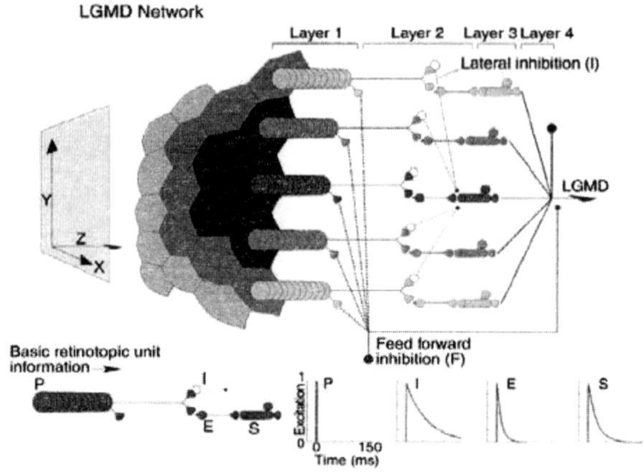

Figure 3. A schematic representation of the neural network model of the LGMD proposed by Rind and Bramwell (1996). In the lower part the response of each processing unit is reported.

3 Experimental Results

Starting from the model discussed in Section 2, a collision detector system for a walking robot has been implemented. A radical change has been applied to the processing units, in fact a biological neuron model has been used, the FitzHugh-Nagumo model (FitzHugh, 1961) reported in the following:

$$\begin{cases} \frac{dv}{dt} = f(v) - w + I_a \\ \frac{dw}{dt} = bv - \gamma w \\ f(v) = v(a-v)(v-1) \end{cases} \quad (1)$$

with $a = 0.5$, γ=2e-3; I_a is the external stimulus and the parameter b is used to modify the system response in order to reproduce the dynamic of the units E, I and S.

This dynamical system when the input is under a given threshold is in a quiescent state, while if the input value increases enough, a spike is emitted and the system cannot be influenced by other external stimuli until the base level is reached.

The proposed looming detector system consists of a MiniCCD (i.e. a 2.4 GHz wireless mini color video transmitting system ALM-2451M) equipped on the robot *Six Black* as shown in Figure 4. The visual system is completely

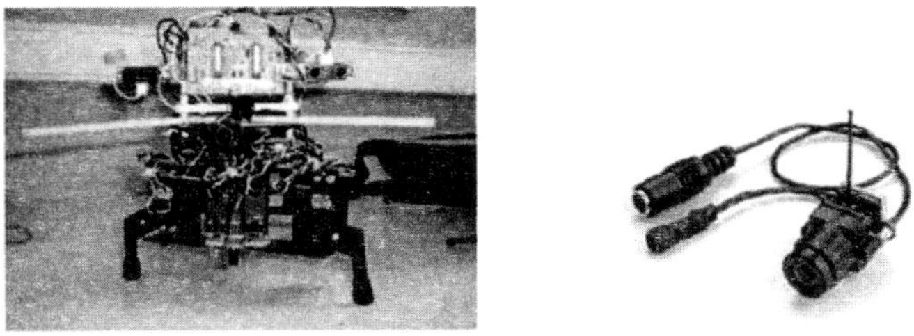

Figure 4. A wireless MiniCCD is equipped on the *Six Black* hexapod.

autonomous and transmits the video stream with a wireless communication system. The video frames acquired with a PC are elaborated and then when a collision is expected, an escape command can be sent to the robot. The results reported here regards an off-line processing, in fact the aim is the methodology validation, however a realtime implementation will be the next step.

The collision detection algorithm can be described by the following step sequence:

1) The images coming from the camera, with a resolution of 320x240 pixels, are converted in Black and White and then a new image is created. This image is the difference between two consecutive frames. This variable represents the change in the level of illumination.

2) An array of 20x15 photoreceptor represents the first layer of the network; each photoreceptor is associated to an image region of 16x16 pixels. Each photoreceptor is activated if the illumination level of the corresponding region is over a given thresholds.

3) The layer 1 output is given to the E and I-units that evolve following the equation (1). Subsequently the S-units collecting the layer 2 output can be processed.

4) The LGMD sums linearly the S-unit outputs and if the value is over the collision threshold, an escape procedure is activated. The collision threshold is chosen in relation to the robot speed and the average speed and dimension of the approaching objects.

5) The distinction between approaching and receding objects is operated through the feedforward inhibition that acts from layer 1 to layer 4 when a high number of P-unit is simultaneously activated. The receding threshold must be chosen adopting the same constrains used for the collision threshold.

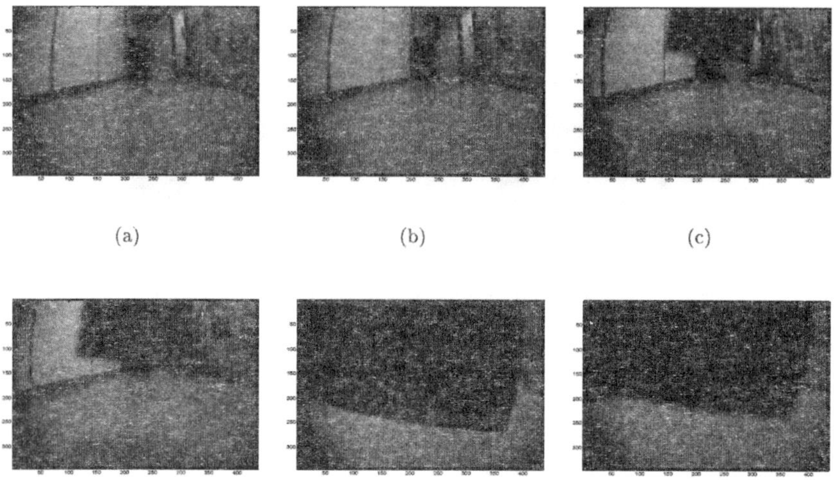

(a) (b) (c)

(d) (e) (f)

Figure 5. Sequence of images transmits by the wireless MiniCCD. The frame rate is $25 frames/s$ and the frame position of each image is: (a) 11, (b) 34, (c) 54, (d) 55, (e) 59, (f) 60.

The whole strategy was validated under different conditions. A typical example of collision detection is here analyzed by using a video acquired from a walking hexapod robot shown in Figure 4 (a). The sequence of images acquired by the MiniCCD is reported in Figure 5.

During the elaboration, the difference between two consecutive images is computed to valuate the changes in the illumination level of the environment that are related to objects in motion (see Figure 6). The acquisition of the information coming from the dynamical environment is assigned to the P-units that generate a response as shown in Figure 7. In this step it is important to make a filter to attenuate the disturbances due to the background that is in motion with respect to the robot and also introduced by the typical pitch and roll oscillations associated to a legged-robot locomotion. The filtering is accomplished setting a moving detection threshold that guides the P-units activation. In the layers 2 and 3 the internal representation of the environment given by the Photoreceptor is elaborated, the state variables of the dynamical systems used in these layers create a short time memory, necessary to identify collision trajectories. The map of S-unit outputs is shown in Figure 8. The LGMD output for the 60 frames

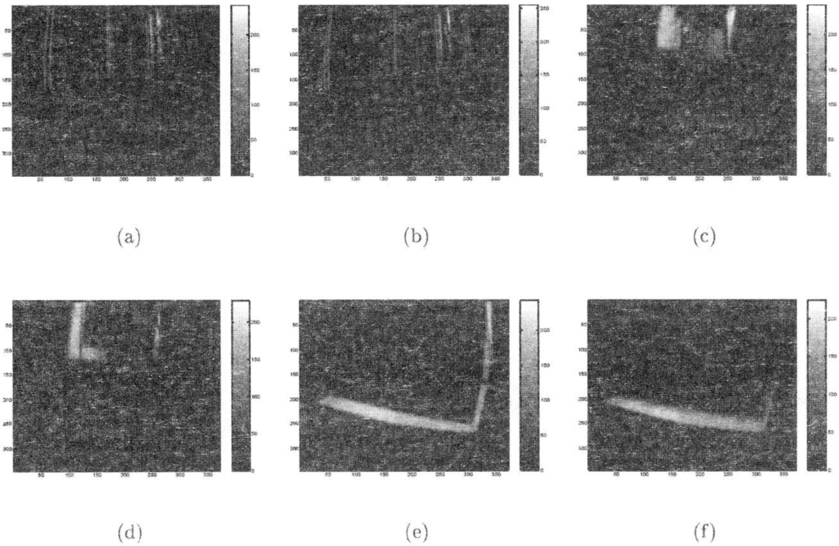

Figure 6. Sequence of illumination level related to the video in Figure 5.

composing the video sequence is reported in Figure 9. The signals has a rapid increase during the last ten frames, when a collision with an object is expected. Within this time interval, the F-unit inhibits the LGMD but the collision threshold, set to 70 in these experimental conditions, has been reached previously triggering an escape behavior. The inhibition is activated when a number of P-units greater than the receding threshold, set to 20 units, is activated in the same frame. The important role of the feedforward inhibition is demonstrated by elaborating the same video but inverting the time direction. In this case the collision threshold is not reached and the LGMD is inhibited by the F-unit. Figure 10 shows the LGMD trend, an object moving away from the robot does not trigger an escape behavior.

A similar experiment has been done in Rind et al. (2003) by using a Khepera robot (bib). The difference between the two experimental applications consists in the dynamical system used to model the elementary units of the LGMD network. A further improvement is given by the fact that the collision detection algorithm has been validated for legged robots that presents several locomotion problem (e.g. pitch and roll oscillations) with respect to a wheeled system.

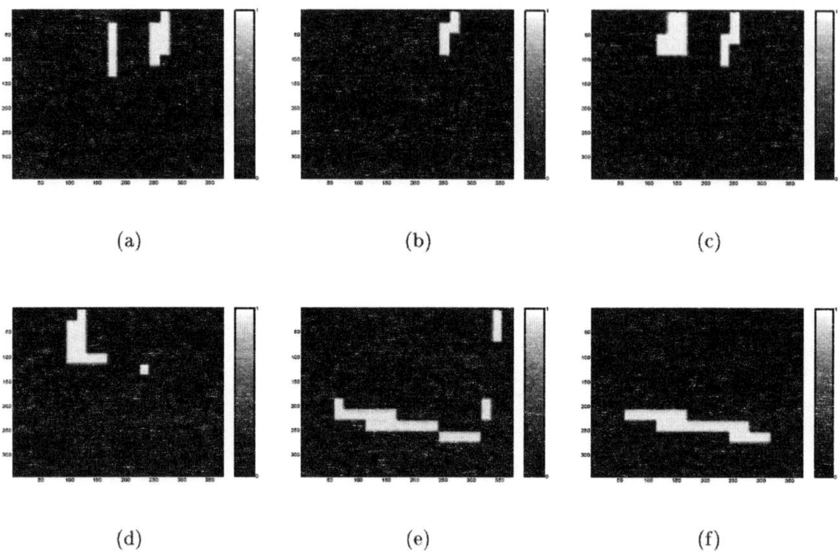

Figure 7. Output of the Photoreceptor layer when the video sequence of Figure 6 is taken into account.

Bibliography

K-team web site. URL `www.k-team.com/robots/khepera/`.

R. FitzHugh. Impulses and physiological states in theoretical models of nerve membrane. *Biophys*, 1, 1961.

N. Franceschini. *Sensors and sensing in biology and engineering*, chapter From fly vision to robot vision: re-construction as a mode of discovery. Barth, Humphrey, Secomb, springerwiennewyork edition, 2003.

F. C. Rind and D. I. Bramwell. Neural network based on the input organization of an identified neuron signaling impending collision. *Journal of Neurophysiology*, 75:967–985, 1996.

F.C. Rind, R.D. Santer, J.M. Blanchard, and P.F.M.J. Verschure. *Sensors and sensing in biology and engineering*, chapter Locust's looming detectors for robot sensors. Barth, Humphrey, Secomb, springerwiennewyork edition, 2003.

G. Sandini and G. Metta. *Sensors and sensing in biology and engineering*, chapter Retina-like sensors: motivations, technology and applications. Barth, Humphrey, Secomb, springerwiennewyork edition, 2003.

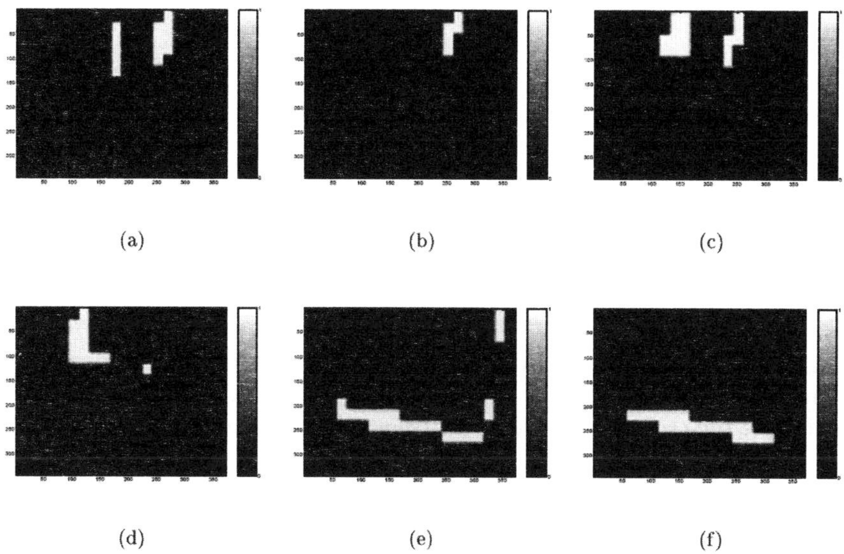

Figure 8. Behavior of the S-units obtained when the layer 1 is activated as in Figure 7

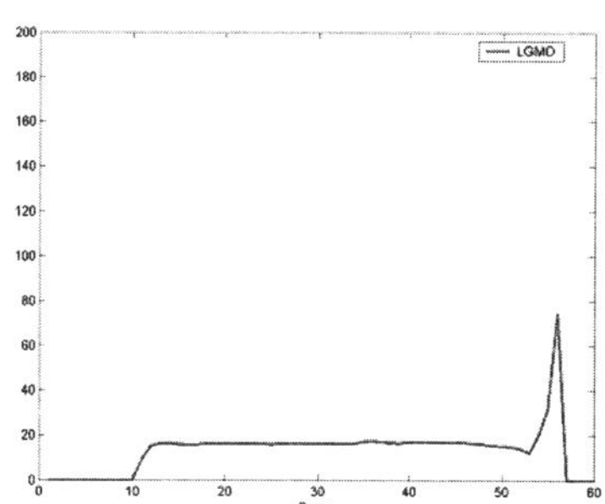

Figure 9. Output of the LGMD network during the robot locomotion. In the last ten frames an escape behavior is triggered by the network.

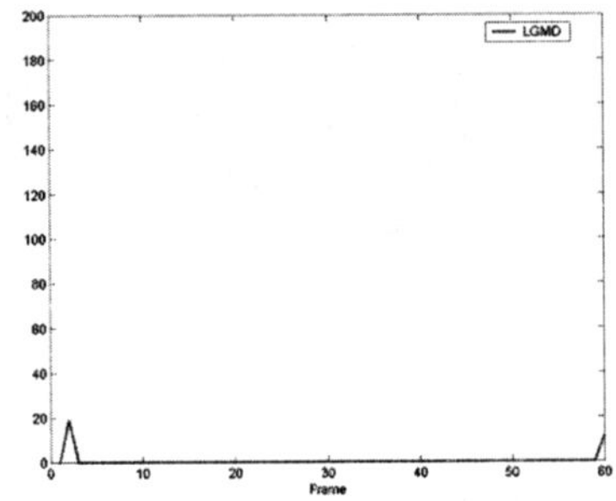

Figure 10. Output of the LGMD network when the robot is walking away from a moving object. The F-unit is activated early and the LGMD is inhibited, the escape response is not triggered as expected.

Hearing: recognition and localization of sound

Paolo Arena, Luigi Fortuna, Mattia Frasca and Luca Patané
Dipartimento di Ingegneria Elettrica Elettronica e dei Sistemi,
Universitá degli Studi di Catania, Catania, Italy

Abstract In this chapter the problem of recognition and localization of a sound source is discussed. The starting point is the solution adopted by the cricket able to identify an auditory stimulus and to reach the sound source. The auditory process has been modelled with a network of neurons with the aim to realize a bio-inspired auditory system for robotic applications.

1 Introduction

The auditory system is an important sensory apparatus for living beings but for a long time the use of this sensorial stimulus in robotics has been set aside with respect to other sensory processes as vision and touch. Moreover the auditory signal can be a rich source of information, in fact several animals take advantages of the sensory system to locate the sound source and interpret the information carried by the sound signal in order to win their daily survival battle. For example the spatial localization of the auditory stimulus is a fundamental skill of the barn owl (i.e. *Tyto Alba*) (Konishi, 1983), a nocturnal predator that adopt an acoustic prey localization during hunting. Another example is the ability of the bat (i.e. *Pipistrellus Chirotteri*) (Walzer et al., 2004) that localizes the preys and the obstacles in the environment emitting a sound and analyzing the reflected echoes. The bat is able to acquire information about the speed of the prey through the Doppler effect, emitting a constant frequency signal. Moreover it can valuate the distance from the target based on the time difference between the emission and the echo reception of a frequency modulated signal.

On the other hand the auditory system is fundamental also for preys. A demonstration is the *noctuid moths* (Payne et al., 1966) that recognize the sound emitted by bats and react in function of the distance from the predator. Mostly important is also the rule of the auditory system in humans since ability to speak is the base of the interpersonal relations.

The sound contains several information that can be summarized in three points:

Who - identity of who send the signal;
What - the message contained in the sound;
Where - the location of the source.

The localization of the sound source (i.e. directional hearing) can be carried out following several different methodologies (Cohen and Gifford, 2001), the most adopted are: the interaural time difference that computes the arrival time difference of the sound in the two ears in function of the source direction and the interaural level difference that computes the intensity difference of the sound perceived by the two ears.

In the following section a model based on the cricket phonotaxis will be presented in details, the neuro-physiological studies have allowed the modelling of the auditory system.

2 Cricket phonotaxis

In nature several strategies are adopted by animals to communicate with others, a well-known example is the cricket song.

Male crickets open and close wings rhythmically, the vibration of the wings generate sound. This song is perceived by the female crickets that is able to: recognize the song and localize the source in order to reach the male.

The cricket song typically consists of multiple groups of 4 syllables, named chirp, emitted with a gap of 20 ms followed by a 340 ms period of silence. The syllable is a 20 ms burst of a 4.7 khz sine wave.

The cricket females are particularly selective to the syllable repetition interval (SRI) in the calling song.

The neural circuit underlying cricket phonotaxis was subject to several studies (Schildberg, 1984; Schildberg et al., 1989). The neural scheme taken into account in this section is a simplified model of the *Gryllus Bimaculatus*'s auditory process, proposed in Webb and Scutt (2000). The simplest model is constituted by only four neurons as schematized in Figure 1.

The auditory neurons (AN) represent the first stage of processing in the cricket's prothoracic ganglion in which a pair of neurons receives direct input from the auditory nerve. The motor neurons (MN) produce an output signal to turn left or right depending on which auditory neuron fires first. Each AN presents a direct excitatory synapse to the corresponding MN and an inhibitory synapse to the opposite MN; the connection scheme is similar to the control system of Braitenberg's vehicles (Braitenberg). The AN-MN excitatory synapses exhibit depression, so the response is optimal for a signal with an appropriate temporal patterning. This basilar model is able to simulate the peculiar characteristics of the cricket behavior in terms of call-

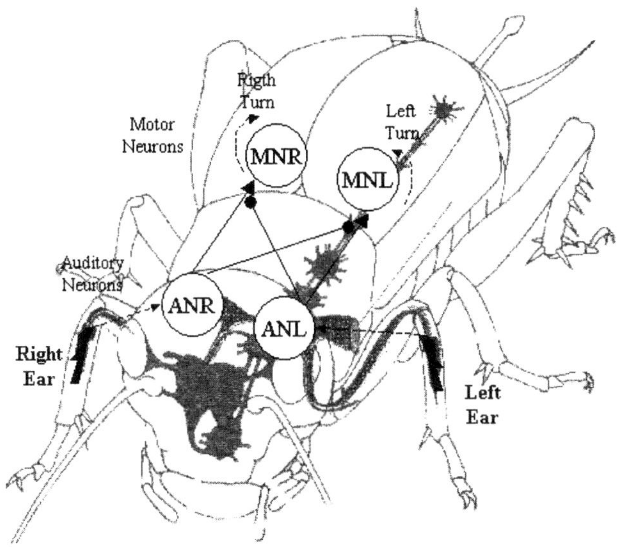

Figure 1. Simplified neural scheme designed to reproduce the cricket phonotaxis.

ing song recognition and sound source localization. In literature this simple model was refined including other elaboration stages as reported in Reeve and Webb (2002). Moreover the problem of sensorimotor integration of different behaviors can be investigated. It is known that the cricket combines its sound response with other sensorimotor activities such as optomotor reflex. For this reason the different sensory systems must be integrated (Webb and Harrison, 2000). Preliminary experiments that validate the feasibility of the application of auditory systems on robots were realized by B. Webb and R. Quinn as illustrated in Horchler et al. (2003); Webb et al. (2003).

In the next section a new model for the implementation of phonotaxis in bio-inspired robots will be proposed. The aim is to formulate a scheme based on nonlinear dynamical systems modelling neuron dynamics, able to localize a specific sound source and to discriminate different emission frequencies that are associated to appetitive or aversive stimuli. For example in the cricket the calling song produces an attractive behavior, while the sound emitted by bats that are predators of the cricket induces an escape

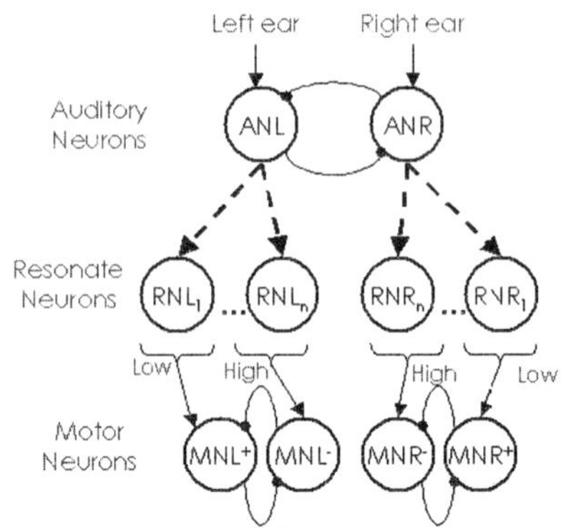

Figure 2. Scheme of the neural model implementing the phonotaxis with the multi-frequency discrimination. Each circle represents a neuron, an arrow indicates an excitatory synapsis, a line with a dot is used for an inhibitory activity while a dashed arrow indicates an excitatory synapsis with depression. The left (labelled L) and right (labelled R) MNs are associated to action of avoidance (MN^-) or approach (MN^+).

behavior (Imaizumi and Pollack, 1999).

3 A phonotaxis neural model with amplitude-frequency clustering

Starting from the basic characteristics of the cricket phonotaxis neural scheme discussed in section 2, a new model has been proposed to include an amplitude-frequency clustering. This model is based on a network of integrate-and-fire and resonate-and-fire neurons and is aimed to control the phonotaxis behavior of a roving robot. A complete scheme of the new model is given in Figure 2, the structure is characterized by three different layers.

In the first layer two auditory neurons are employed to elaborate the signals coming from the sensory system (e.g. microphones equipped on the robot). The ANs are modelled with the leaky integrate-and-fire (IF) neuron (Koch; Izhikevich, 2000), the equation is the following:

$$\begin{cases} \dot{x} = -kx + \sum I_{syn} + I_{dis}y \\ y(t+1) = \begin{cases} 1 & if x > 1 \\ 0 & if x < 0 \\ y(t) & otherwise \end{cases} \end{cases} \tag{1}$$

where x is the membrane potential, y is the neuron output (i.e. a train of spikes), $\sum I_{syn}$ is the sum of the afferent synaptic currents, I_{dis} is the discharge current and k weights the leaky discharge. The IF neurons ANL and ANR are mutually inhibited so that there is a winner-take-all-effect. The winning neuron is the one associated with the highest amplitude signal; in this way, the two neurons discriminate the amplitude difference between the two microphones.

Therefore, the first layer is devoted to the localization of the sound source through the identification of the direction. The second layer consists of an array of resonate-and-fire neurons (RN) (Izhikevich, 2001) tuned at different frequencies. This layer is devoted to frequency clustering.

The first layer is connected to the second layer through depression synapses that have been designed in order to associate a single spike to a burst coming from the corresponding AN.

The equation of the depression synapsis is reported in following:

$$\begin{cases} \dot{Dep} = -a_1 Dep + a_2 AN_{out} \\ RN_{in} = a_3 AN_{out} - Dep \end{cases} \tag{2}$$

where a_i are parameters and Dep is the depression effect.

In the second layer the Morris-Lecar model (Morris and Lecar, 1981; Hoppensteadt and Izhikevich, 2002) is adopted to implement the resonate neurons, the equations of the model are the following:

$$\begin{cases} \dot{V} = k_f[I + g_l(V_t - V) + g_k w(V_k - V) + g_{Ca} m_\infty(V)(V_{Ca} - V)] \\ \dot{\omega} = k_f[\lambda(V)(\omega_\infty(V) - \omega)] \end{cases} \tag{3}$$

where

$$m_\infty(V) = \frac{1}{2}(1 + \tanh\frac{V - V_1}{V_2})$$
$$\omega_\infty(V) = \frac{1}{2}(1 + \tanh\frac{V - V_3}{V_4})$$
$$\lambda(V) = \frac{1}{3}\cosh\frac{V - V_3}{2V_4}$$

V and ω are the state variables of the system, I is the input and V_i, V_t, g_l, g_k, g_{Ca} and k_f are parameters of the model.

The Morris-Lecar neuron emits spikes only if stimulated with a train of spikes showing a particular inter-spiking frequency; the neuron characteristic frequency can be tuned acting on the parameter k_f. In Figure 3 the system behavior is shown. The membrane potential (V) shows damped

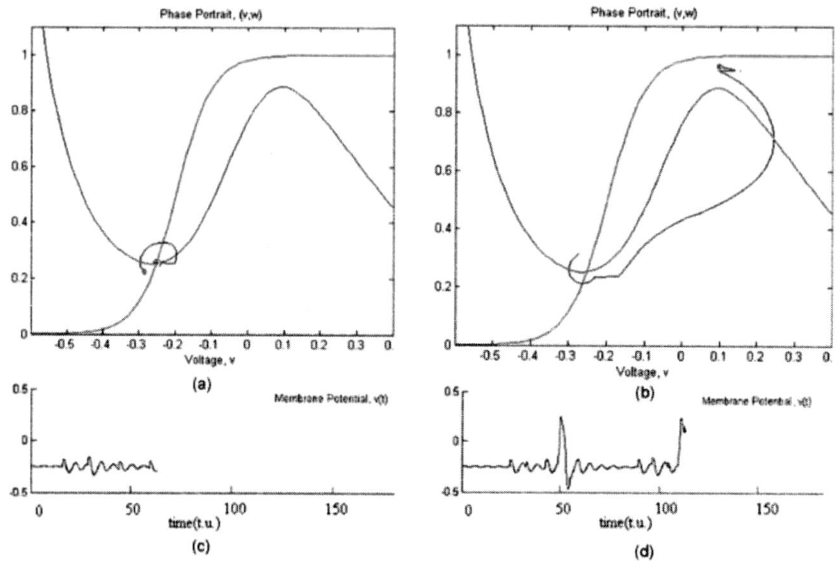

Figure 3. Behavior of the Morris-Lecar model: (a)-(b) phase portrait of the state variable V and ω, in black is shown the evolution of the system, the nullclines $\dot{V} = 0$ and $\dot{\omega} = 0$ are also shown; (c)-(d) time evolution of the membrane voltage (V). When the frequency of the stimulus is uncorrect the system shows damped oscillations (a)-(c) while when the correct stimulus is applied, the system emits spikes (b)-(d).

oscillations around the stable equilibrium point ((V, ω)=(0, 0)), when the frequency of the external stimulus corresponds to the neuron characteristic frequency, the system is forced toward a limit cycle and emits spikes.

The third layer of the control scheme defines the actions that the robot will accomplish, based on the amplitude-frequency clustering. It consists of motor-neurons, modelled as integrate-and-fire neurons.

The system is able to recognize a wide frequency range; in the proposed scheme a low frequency signal produces an attractive behavior, the MN^+ is activated; while a high frequency signal triggers an aversive action, the MN^- is activated.

To obtain more complex behaviors each RN can be associated to a specific network of MNs that will be dedicated to accomplish a particular motor activity.

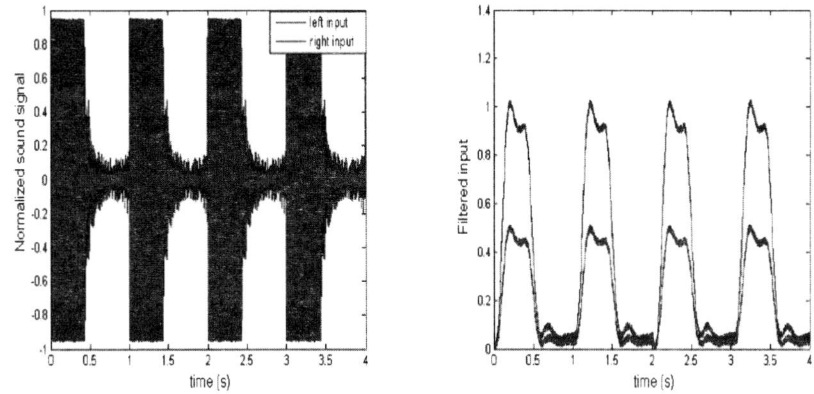

(a) (b)

Figure 4. (a) Sound signals acquired from the two microphones equipped on a roving robot. The data are scaled in [-1, 1]. (b) The sound is preprocessed with a 4-th order Buttherworth filter, the signals are scaled in [0, 1].

4 Simulation results

The model for amplitude-frequency clustering shows different behaviors associated with different characteristics of the input signals (amplitude and frequency). The behavior implemented on a roving robot is similar to the cricket behavior, where some frequencies are associated with the calling song of male crickets, while other ones indicate the presence of predators. Therefore, the whole model for auditory perception is devoted to control different responses (attractive or repulsive) depending on the input characteristics.

The results of the application of this control scheme for the auditory perception in a roving robot, is reported in following.

The robot used to test the performance of the amplitude-frequency clustering, is a dual drive Lego roving robot. The robot was equipped with 2 microphones and the auditory signals are processed with a PC. The song generated with an audio reproducer is acquired by the microphones and elaborated with a 4-th order Buttherworth filter (Cichocki and Belouchrani, 2001). It is a bandpass filter used to eliminate the high frequency oscilla-

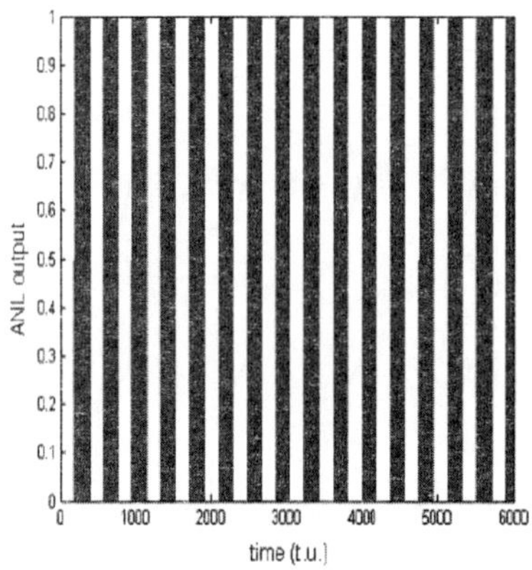

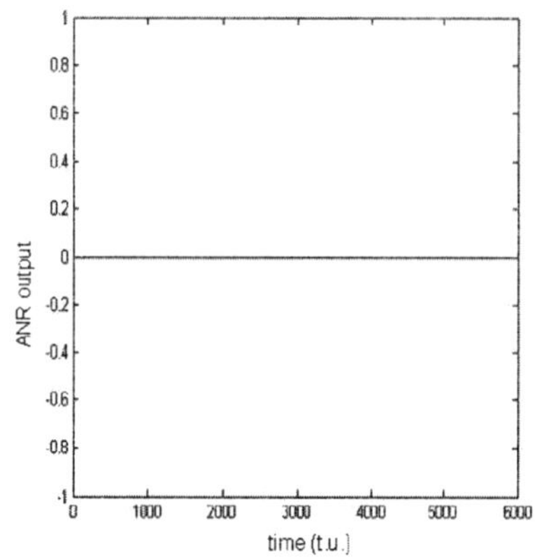

Figure 5. Output of the two auditory neurons: (a) ANL, (b) ANR. The ANL wins the competition.

tions in order to process only the characteristics of the input signals relevant to the phonotaxis problem as the SRI. In Figure 4 the microphone and filter outputs are shown.

The auditory sensory stimuli are processed by the ANs, in this case the ANL wins the competition and emits spikes as shown in Figure 5. Therefore, in the first layer the direction of the incoming sound is identified.

Subsequently the AN output modulated by the depression synapses, is elaborated by the 2-nd layer. The output of the RNs is given in Figure 6. As can be noticed only one RN is correctly stimulated by the input frequency.

In the final step the information acquired from the auditory perception model is applied to modulate the robot behavior. As shown in Figure 7, the frequency of the external stimulus is associated to a left avoidance response in order to escape from possible predators.

Moreover the phonotaxis control scheme can be easily realized in hardware and equipped on autonomous roving robots. The sensing capabilities can be also improved including several microphones that can discriminate the location of the sound source in shortest time and with high precision.

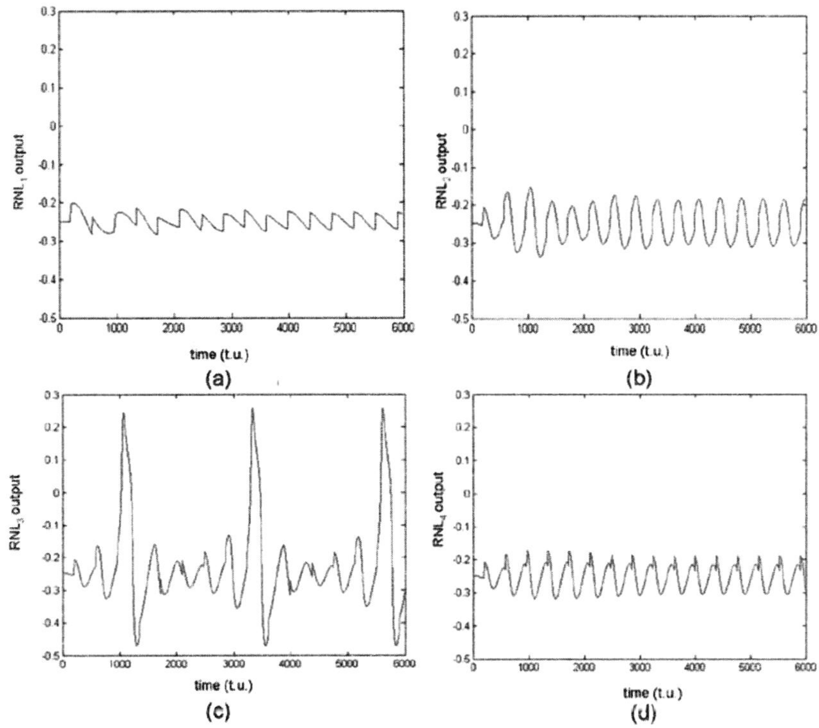

Figure 6. Trends of the membrane potential of the RNs coupled with the ANL. The RNs of the right side are not stimulated by the ANR. Only the neuron (c) resonates at the frequency of the input signal.

Bibliography

V. Braitenberg. *Vehicles: experiments in synthetic psycology.* MIT Press: Cambridge, MA, 1984.

A. Cichocki and A. Belouchrani. source separation of temporally correlated source using bank of band pass filters. In *3rd International Conference on Indipendent component analysis and blind signal separation*, December 2001.

Y.E. Cohen and G.W. Gifford. Neural mechanisms of sound localization. In *Proceedings from the 8th Annual Research Symposium of the SHHH*, 2001.

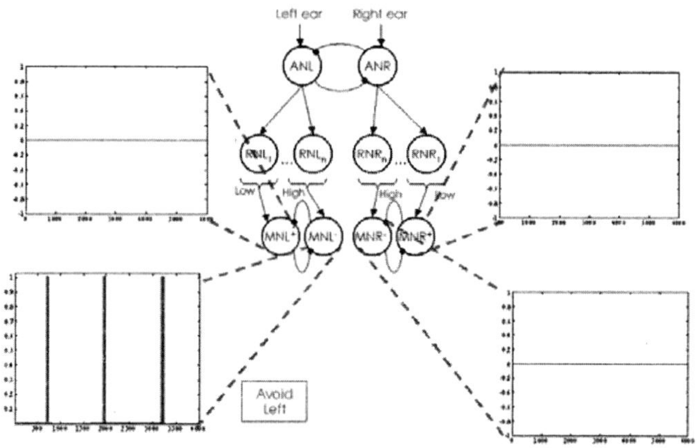

Figure 7. Robot response modulated by the external stimulus, a left avoidance action is triggered to escape from a possible predator.

F.C. Hoppensteadt and E.M. Izhikevich. *Brain Theory and Neural Networks*, volume 181–186. Arbib MA, mit press, cambridge edition, 2002.

A. Horchler, R. Reeve, B. Webb, and R. Quinn. Robot phonotaxis in the wild: A biologically inspired approach to outdoor sound localization. In *Proceedings of ICAR*, 2003.

Kazuo Imaizumi and Gerald S. Pollack. Neural coding of sound frequency by cricket auditory receptors. *The Journal of Neuroscience*, 19(4):1508–1516, February 1999.

E.M. Izhikevich. Neural excitability, spiking, and bursting. *International Journal of Bifurcation and Chaos*, 10:1171–1266, 2000.

E.M. Izhikevich. Resonate-and-fire neurons. *Neural Networks*, 14:883–894, 2001.

C. Koch. *Biophysics of Computation, Information Processing in Single Neurons.* Oxford University Press, 1999.

M. Konishi. *Neuroethology and Behavioral Physiology*, chapter Neuroethology of acustic prey localization in the barn owl. F. Huber and H. Markl, springer edition, 1983.

C. Morris and H. Lecar. Voltage oscillations in the barnacle giant muscle fiber. *Biophysical Journal*, 35:193–213, 1981.

R. Payne, K.D. Roeder, and J. Wallmann. Directional sensitivity of the ears of noctuid moths. *J Exp Biol*, 44:17–31, 1966.

R. Reeve and B. Webb. New neural circuits for robot phonotaxis. *Philosophical Transactions of the Royal Society A*, 361:2245–2266, 2002.

K. Schildberg. Temporal selectivity of identified auditory interneurons in the cricket brain. *J Comp Physiol*, 155:171–185, 1984.

K. Schildberg, F. Huber, and Wohlers D. *Cricket behavior and neuroobiology*, chapter Central auditory pathway:neural correlates of phonotaxis behavior. Huber, F. and Moore, T.E. and Loher W., cornell university press edition, 1989.

V.A. Walzer, H. Peremans, and J.C.T. Hallam. One tone, two ears, three dimensions: A robotic investigation of pinnae movements used by rhinolophid and hipposiderid bats. *J Acoust Soc*, 2004.

B. Webb and R. Harrison. Integrating sensorimotor systems in a robot model of cricket behaviour. In *Proceedings of the Society of Photo-Optical Instrumentation Engineers (SPIE)*, volume 4196, pages 113–124, 2000.

B. Webb and T. Scutt. A simple latency dependent spiking neuron model of cricket phonotaxis. *Biological Cybernetics*, 82(3):247–269, 2000.

B. Webb, R. Reeve, Horchler, and R. Quinn. Testing a model of cricket phonotaxis on an outdoor robot platform. In *Proceedings of TIMR03*, 2003.

Perception and robot behavior

Paolo Arena, Davide Lombardo and Luca Patané
Dipartimento di Ingegneria Elettrica Elettronica e dei Sistemi,
Universitá degli Studi di Catania, Catania, Italy

Abstract The phenomenon of perception is a complex process including several distinct elements that will be described in this chapter. The principles of locomotion and sensing are here re-elaborated from the agent point of view. The sensory information is processed to perceive the environment in order to generate an action useful to accomplish a particular task. Some psychological principles are discussed and the paradigm of action-oriented perception is taken into account for the realization of a perceiving robot.

1 Introduction

In psychology and cognitive sciences, perception is the process of acquiring, interpreting, selecting, and organizing sensory information. In the last decades, the opportunity to apply the principles of human sciences in engineering fields produced a radical innovation and nowadays perception represents a strategic area in robotics research.

The development of intelligent, flexible robots that can adapt to their environment will require sophisticated control systems. In order to operate in an intelligent manner, a robot needs information about itself and its environment. There are several interesting research activities about the interaction of the various knowledge sources and about the consequences of active perception on robotic planning systems.

Perception in robotic systems generally requires data to be detected using sensors and then processed into a form which can be used effectively. Some sensors are based on principles similar to the human senses, for example vision systems, sound and touch sensors. However, sensors on robots are not limited to simply emulating human senses, and robots are commonly equipped with laser range finders, infra-red proximity detectors and GPS systems.

The processing of sensor information can also be done in a myriad of ways, and is usually a fairly intensive activity requiring a significant amount of computing power. Research activities in robotics are greatly involved in developing successful and efficient techniques for processing data acquired from sensory systems.

Most robots are equipped with a number of different sensors, which enable them to obtain a variety of information about the state of themselves and the environmental situations. This is important because no single sensor is infallible, and generally data from the various sources must be combined in some way to arrive at a meaningful, reliable perception of the robot reality.

Traditional perception is based on the construction of an intention-free model of the world (Arkin, 1998); more recent approaches, relying on behavior-based systems, take into account the motor system requirements. In the new view, perception is considered as a synergistic process deeply intertwined with the agent's cognitive and locomotion system. Perception without the context of action is meaningless.

The new perceptual paradigm includes:

- Action-oriented perception: perceptual processing is devoted to produce motor activities.
- Expectation-based perception: knowledge of the world can constrain the current interpretation of the external environment.
- Active perception: the agent can use motor control to enhance perceptual processing.

Perception can be considered as a top-down process; the underlying principle is that perception is predicted on the needs of action. Only the perception necessary for a particular task needs to be extracted from the environment (Arkin, 1998). Therefore the world's view is mediated by the agent's intentions.

In fact in nature, animal experience begins with information about the world that flows in through sensory organs and sensory pathways. The behavior of an animal depends on how it combines that information with its internal states in order to do something (Cruse, 2003).

In the last years, neuroscientists have gained insight in the neural mechanisms responsible for learning and adaptation in biological systems. The principal elements constituting the perception process can be distinguished in sensor systems, learning mechanisms and memory structures.

The sensor system is a fundamental stage in the perception processes. It represents the interface between the external world and the animal internal world. A complete sensory fusion is the first perception step: all sensors cooperate with each other in determining a valid percept.

The basis of the modern Neurobiology, based on the neural mechanisms involved in the learning processes, was proposed by two psychologists: Donald Hebb and Jerzy Konorski to around the end of the Forties . They proposed the idea that learning and memory emerge from physiological modifications in the nervous circuits.

In the 1949 Hebb proposed the well-known Hebb's rule (Hebb): "When an axon of cell A ... excites cell B and repeatedly or persistently takes part in firing it, some growth process or metabolic change takes place in one or both cells so that A's efficiency as one of the cells firing B is increased". In other words, when a cell is active, its synaptic connections are more efficient, producing a temporary increase in the cell excitability or a structural change in the synapsis.

Learning is an essential part of an intelligent system. Learning mechanisms are adaptive changes in behavior caused by experience. Different categories of learning and memory could be identified (Shepherd).

Simple learning: Nerve cells have a number of properties that change during or after stimulation. Different plastic changes occur at synapses as a consequence of activity. Two mechanisms can be distinguished: habituation and sensitization. Habituation is a decrease in the behavioral response that occurs during repeated presentation of a stimulus otherwise sensitization is the enhancement of a reflex response by the introduction of a strong stimulus.

Associative learning: It occurs when an animal makes a connection through its behavioral response between a neutral stimulus and a second stimulus that is either a reward or punishment. The classes of learning based on this mechanism are: classical conditioning, operant conditioning and aversion learning.

Memory is necessary for learning and may be defined as the storage and recall of previous experiences (Shepherd). The memory mechanisms can operate over time scale from few seconds, *working memory*, to years, *associative memory*. Either contain facts and events but the working memory has a duration from millisecond to second for a moment-to-moment utilization of information, otherwise the associative memory has a duration of many years with the aim of acquisition of knowledge and experience. As noted by Patricia Goldman Rakic (1991): "if associative memory is the process by which stimuli and events acquire permanent meaning, working memory is the process for the proper utilization of acquired knowledge".

In this chapter, the various classes of perception will be adopted to define efficient algorithms for robot navigation.

2 Perception in robotics

The results achieved by the international community during the last years in the application of psychological principles in robotics and in particular for the realization of active perception are encouraging and open the way to further developments.

In this chapter the perception paradigms will be applied to artificial systems following an increasing level of abstraction and complexity. We will introduce a robot in which perception is implemented at several levels.

In particular the foraging task is examined from the point of view of a perceptual machine. The aim of a robot involved in a foraging task is to collect several targets located in the field and to avoid obstacles. To make the robot able to accomplish the given task a hierarchical control system (as the scheme developed by Verschure, see Section 3.1) has been introduced. Different levels of behavior can be distinguished in this perceiving robot (Selfridge and Franklin, 1990).

The first level that constitutes the basis of a perceptual process is the *reactive layer*. The capability of linking low level stimuli (i.e. touch sensors for obstacle collisions or target retriever) with actions through pre-wired connections. Applying this control level the robot shows some basic behaviors, a number of fixed actions that will be given by default, to simulate inherited reactions. In particular for the foraging task the robot is able to change the moving direction generating a repulsive response when a collision occurs in order to avoid the obstacle or an attractive response when a target is identified along the path. Moreover the system capabilities can be improved introducing a further control level. The *adaptive layer* is hierarchically located over the reactive layer. It gives to the structure new capabilities that improve the robot behavior. It is able to learn how to use new sensors that are related to the pre-wired sensors. Therefore the system can learn how to avoid collisions before the contact through range finder sensors. However at the reached level of complexity, the foraging is accidental, the robot executes explorative movements until it reaches a target. In order to shift the robot from a reactive towards a deliberative behavior, it is necessary to introduce a new control layer in which useful information can be memorized. When an important event occurs (e.g. a target is found), the sensing/action pairs that have been carried out to attain this objective are stored in memory so that they can be reused in similar circumstances. The perceptual scheme described above represents the Distributed Adaptive Control scheme developed in Verschure et al. (2003) and realized with static artificial neural networks mapping sensing into actions. An aspect that is not faced with this perception scheme, is the search of the optimal solution of the task. Proceeding in the abstraction mechanism, it is possible to introduce a new layer that will be used to guide the foraging according to optimal criteria based on the path length and the number of targets found. The approach that will be discussed in Section 4 is based on the Neisser's Perceptual Cycle. The algorithm creates maps of the explored field with information about the distance between the identified objects in relation to the target. The cognitive map represents a reward function that will be used during the robot navigation.

3 Distribute adaptive control

3.1 DAC5

The application of perceptual mechanisms in robotics is a new frontier that involves several research groups of the international community. An innovative scheme based on action oriented perception, realized at the Institute of Neuroinformatics, Zurich, has been applied to our perceptual robot in order to deal with a foraging task. The perceptual scheme, named Distributed Adaptive Control (DAC5), is based on the neural model of classical and operant conditioning (Verschure et al., 2003; Verschure and Althaus, 2003).

Classical conditioning, initially studied by Pavlov, assumes that unconditioned stimuli (US) automatically generate an unconditioned response (UR). The relationship is defined genetically and is important to ensure the survival of the agent. For example in Pavlov's studies, the sight of food (US) produces dog's salivation (UR). Pavlov noticed that an association with conditioned stimulus (CS) could be developed. In the dog's case, if a bell rung repeatedly in association with the sight of food, over time the bell alone was sufficient to induce salivation.

Summarizing two different classes of stimuli can be distinguished:

- Unconditioned Stimuli (US) representing basic reflex actions where low-complexity sensory events trigger simple actions, Unconditioned Responses (UR);
- Conditioned Stimuli (CS) are initially neutral stimuli, but become able to trigger Conditioned Responses (CR) due to their correlation with motivational stimuli (US).

Presentation of an US causes an automatic response UR; after several simultaneous presentations of US and CS, the presentation of the CS alone will trigger a response (CR) similar to the UR.

Starting from these neurobiological principles a perceptual algorithm has been proposed in Verschure et al. (2003, 1992). The DAC5 architec-

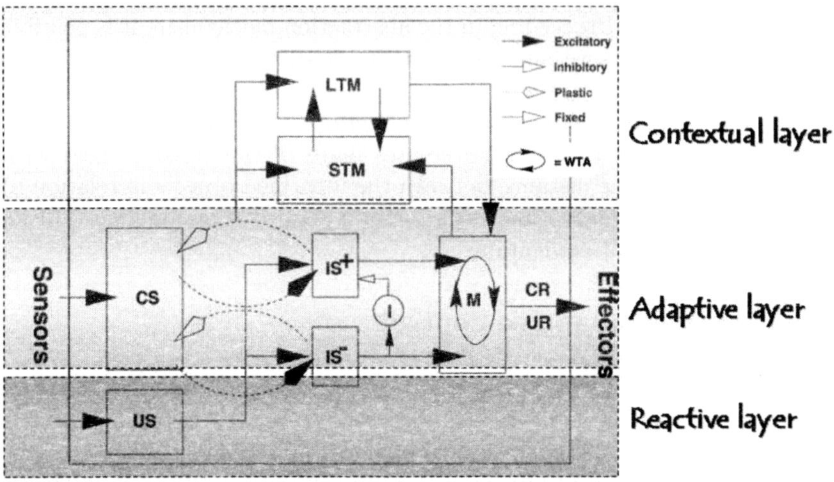

Figure 1. DAC5 Scheme.

ture scheme, shown in Figure 1, consists of three control layers: reactive, adaptive and contextual layer.

Reactive and Adaptive layer The reactive control layer implements a basic level of competence based on pre-wired reflexive relationships between simple sensory events (US) and actions (UR).

The adaptive control is used to develop a representation of complex sensory events (CS) through the relation between simple events (US).

An US activates specific populations of neurons reflecting an internal state (IS); the relation could be aversive and appetitive. IS cells will activate specific reflexive actions (UR) categorized in avoidance and approach. Conflict resolution in action selection is resolved with an inhibitory unit that privileges avoidance with respect to approach. The CS representations are obtained, during the learning phase, by modifying the connection weights between the CS and IS populations.

Contextual layer The contextual layer supports the formation of more complex representations of CS and CR events expressing their relationship in time.

Important events are stored in the short-term memory (STM), while the content of STM is stored in a long-term memory (LTM) when a collision occurs or a target is found. This layer can be activated when the internal representations of CS constructed in the adaptive layer are stable. During the learning phase, the CS events and the corresponding responses are memorized in the STM realized with a ring buffer (i.e. a FIFO queue). When the robot reaches a target or makes a collision, the STM is downloaded in the LTM. During the robot exploration the chain of actions stored in the LTM is compared to the sensory events generated by the adaptive layer. The LTM elements similar to the CS prototype compete for the control of behavior. The contextual layer may control the robot only if the reactive layer is not active.

3.2 Simulation results

The DAC5 control system previously described represents a fundamental example of application of perceptual mechanisms in robotics. It is a hierarchical scheme based on action-oriented perception and can be used as a benchmark.

Therefore a simulation tool was designed and implemented in Java; the idea is to realize a framework in which different perceptual paradigms can be validated when applied on simulated robots. In the tool several items can be defined: the robot structure, characterized by specific dimensions and equipped with different types of sensors; the environment that includes objects that interact with the robot. The framework has been used to test the perception strategy proposed in Verschure et al. (2003); Verschure and Althaus (2003).

A simulated robot is faced with a foraging task where collisions had to be minimized while the number of targets found had to be maximized. The simulated environment consists of a two dimension arena with multiple obstacles and targets. The robot has a circular shape, it is equipped with three types of sensors: 37 collision sensors (US^-), 37 range finder sensors (CS) and 2 target sensors (US^+). The collision and range finder sensors are distributed in the front of the robot while the target sensors are positioned on the left and right side.

The robot is able to execute five predefined actions:

- Avoid_left (right) - go back and rotate 9° on the left (right).
- Approach_left (right) - rotate 9° on the left (right) and go forward.
- Exploration/Approach_forward - go forward.

The parameters used to valuate the performance of the control levels are: number of collisions in relation to the number of target detections, the number of avoidance movements stimulated by CS and behavioral entropy

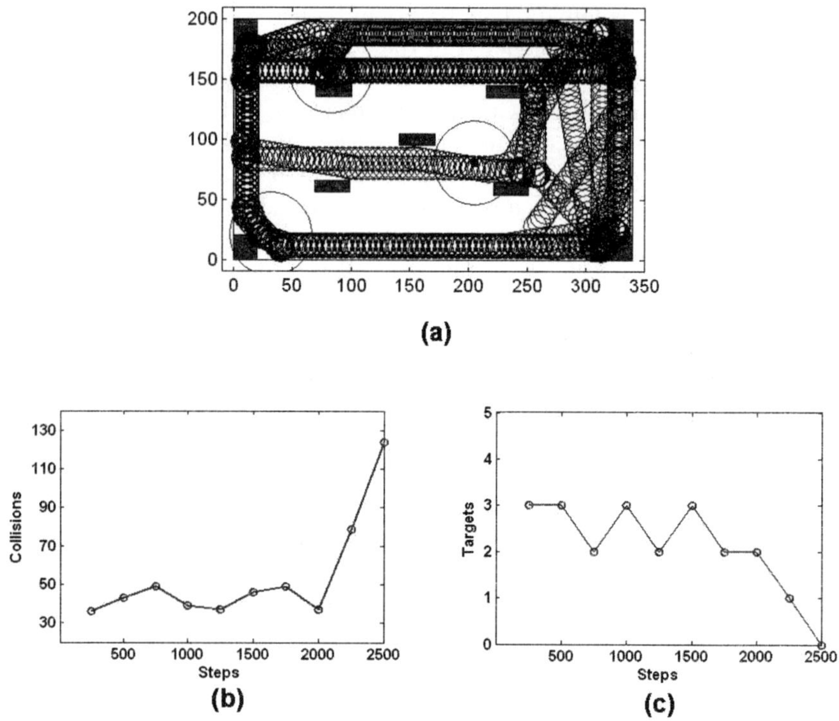

Figure 2. Simulation results obtained with the Reactive layer. (a) Trajectory of the robot covered in 2500 simulation units, (b) number of collisions and (c) number of targets found. The events are calculated in windows of 250 simulation units.

(Verschure et al., 2003) that is the capability to learn stable trajectories that can be repeated.

When only the reactive layer is active, the robot explores the environment colliding with obstacles and reaching the targets when the target-finder sensor is within the target perception range. The results obtained adopting this control layer are reported in Figure 2. The number of collisions is very high and in the last period increases rapidly, in fact the robot is not able to escape from a deadlock and for this reason the number of targets found decreases to zero.

The introduction of the Adaptive layer increases the robot capability; after some collisions the robot has learned to act so as to find the targets

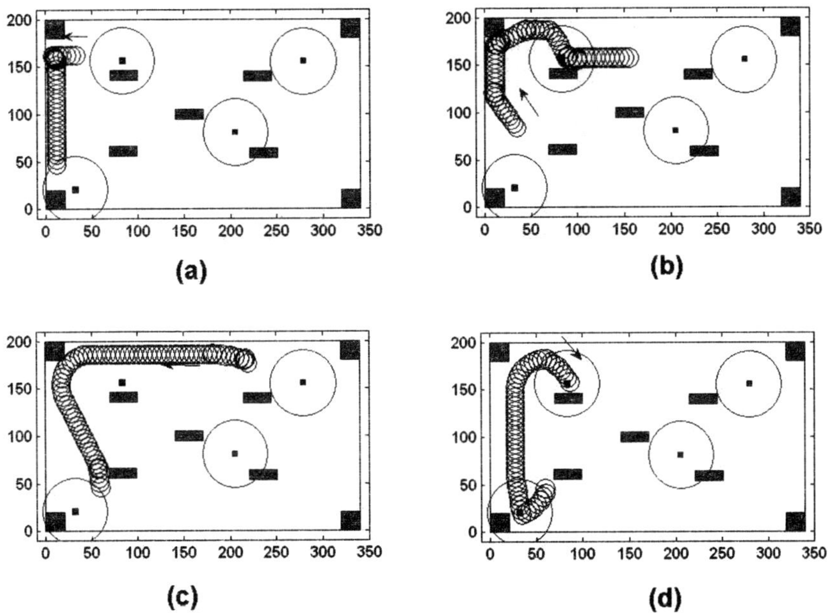

Figure 3. Robot simulation when the Reactive and Adaptive layer are active. (a) Steps 1-50, a collision occurs; (b) steps 101-200, a target is founded after several collisions; (c) steps 401-500, the robot avoids an obstacle through a CR, one of the target is not active because it is the last target founded; (d) steps 601-700, the trajectory is smoothed and two targets are found.

avoiding obstacle collisions. A sequence of actions carried out by the robot is shown in Figure 3.

The complete path and other simulation results are shown in Figure 4. The number of targets found after an initial transient, reach a value of 3-4 in windows of 250 simulation units against a value of 2-3 reached when only the reactive layer is active. As it can be noticed in Figure 4 (b) after 600 steps the US produced by physical collisions are absent while the avoid actions stimulated by CS increase with the learning. The behavioral entropy is H=6.64 and can be compared with the value of H=7.27 obtained with the reactive layer before the deadlock. Therefore the introduction of the adaptive layer increases the trajectory stability of the robot and also increases the number of collected targets.

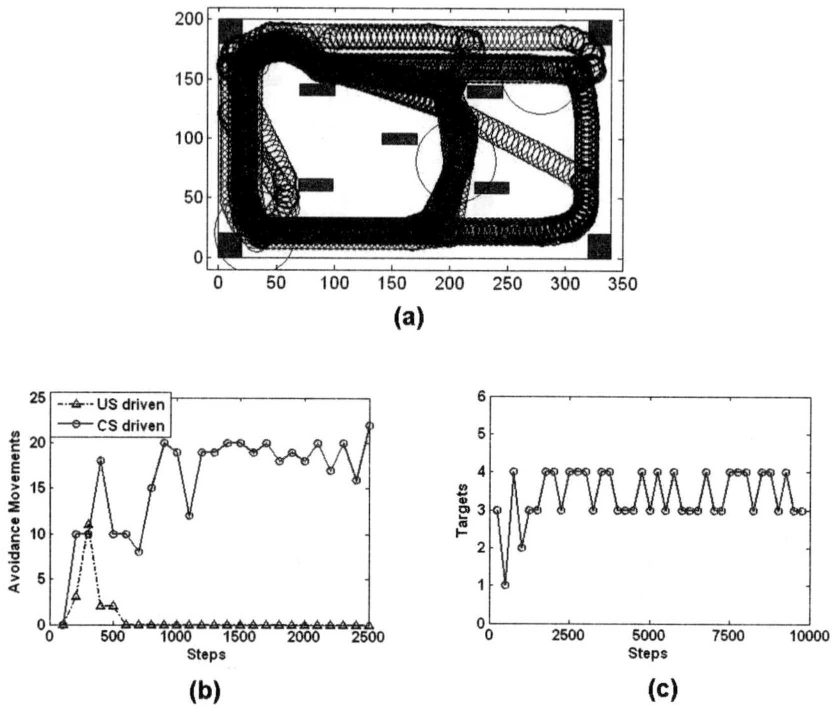

Figure 4. Simulation results obtained when the Reactive and Adaptive layer are active. (a) Trajectory of the robot covered in 2500 simulation steps, (b) number of collisions avoided by CS and US, (c) number of targets found.

The last simulation includes the introduction of the contextual layer. The application of information stored in memory can induce the robot to a sub-optimal solution for the foraging task. The results reported in Figure 5 underline this problem. The robot reaches a target rate next to 3, less than the value of 3-4 obtained previously; the reason is the length of the path learned that is not the shortest one; the system is bound in a local minimum. However the order introduced by the contextual layer reduces furthermore the behavioral entropy to a value H=6.46.

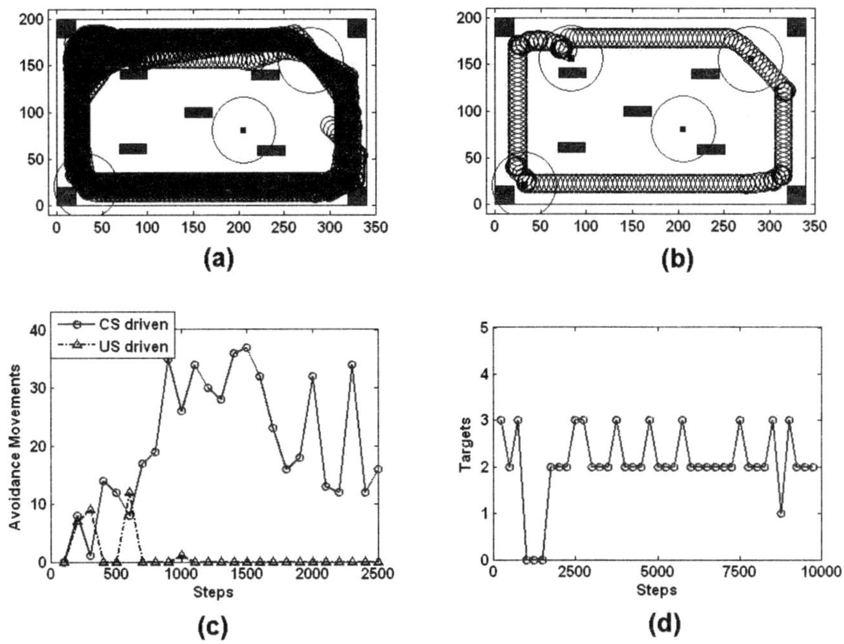

Figure 5. Simulation results obtained when the Contextual layer is active. (a) Trajectory of the robot covered in 2500 simulation steps, (b) trajectory learned and stored in memory; (c) number of collisions avoided by CS and US, (d) number of targets found.

4 Cognitive maps for robot navigation

4.1 Cognitive maps

In general two primary information sources could be used to perceive the external world in accurate way: the sensory input immediately available and the previous experience or acquired knowledge memorized in the brain.

The terms bottom-up o data-driven-processing indicate the perceptual processes influenced only by the sensory input, while the terms top-down or conceptually-driven-processing describe the processes based on past experiences or on the contextual information, related to the environment in which the perception occurs. The top-down processing is also known as expectation.

It is evident that perception includes either bottom-up and top-down processes, then the aim is to understand how they influence perception and to estimate their interaction links.

From this debate several theories have emerged: Gregory in 1970 proposed the Constructivist Theory based on a top-down processing (Gregory, 1980). In the same period Gibson formulated the Direct Perception Theory. Following the bottom-up approach he asserted that there is much more information potentially available in the sensory stimuli then can be believed (Gibson, a,b).

Affordance is a concept introduced in Gibson (b): affordances are the potentialities for action inherent in a object or scene, important cues in the environment that aid perception.

One theory that explains how top-down and bottom-up processes may be seen as interacting with each other to produce the best interpretation of the stimulus was proposed by Neisser: it is known as the Perceptual Cycle. The Perceptual Cycle is constituted of schemes that contain the knowledge acquired in past experiences. The schemes, named also cognitive maps, are dynamical structures that guide the environment exploration towards relevant stimuli. The subject moving in the environment acquires sensory information and compares it with the scheme elements that will be modified.

The bottom-up-processing is represented from the sampling of the available environment information that can modify the perception scheme. The top-down approach is included in terms that the schemes influence the perception processes.

A schematic representation of the Neisser's Perceptual Cycle is given in Figure 6.

4.2 Application to robot navigation

The robot navigation is a complex problem that may be faced with different techniques in relation to the measurement system available on the robot. Adopting the strategy of the Perceptual Cycle previously introduced, a methodology for the localization of a target through landmarks is proposed. The robot exploring the environment constructs an orientation scheme identifying the objects in the scene through features extraction. When the target is found, a given number of objects previously identified is stored in memory, associating to each one a reward value that represents the relation with the target in terms of distance. Therefore during the environment exploration the robot selects the interesting objects extracting the available information; the perception of the external environment modifies the internal schemes (i.e. Cognitive map) that create a distance graph between all the acquired objects in relation to the robot task that is to find

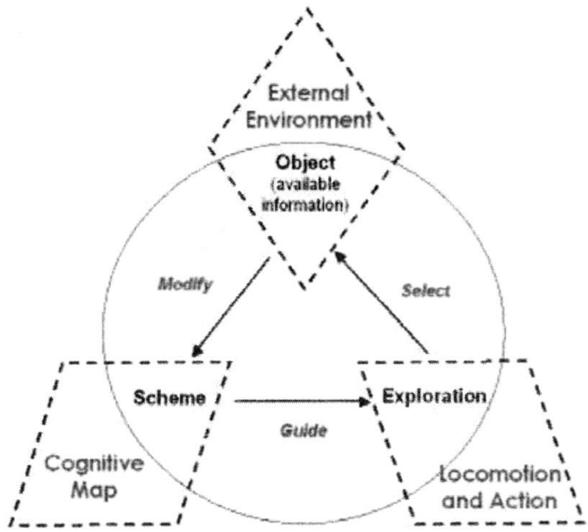

Figure 6. The Perceptual Cycle proposed by Neisser.

and reach a target. The continuous development of the schemes is essential in a dynamic changing environment and the information stored in memory guides the successive exploration closing the Neisser's Perceptual Cycle as in Figure 6.

The methodology has been validated carrying out several experiments on a Lego robot. The system is equipped with a webcam and controlled by PC; the environment contains multiple objects with the same shape but different color. The robot, acquiring the images from the webcam, applies a segmentation procedure in order to distinguish the objects in the scene. A set of features are extracted from each object and all the information are stored in the working memory (STM) realized as a ring buffer. When the target is localized, the information stored in the STM are transferred into the long-term memory (LTM) and a reward value is assigned to each object in relation to the distance from the target. The elements stored in the LTM represent landmarks and in a second exploration the robot is guided in the field trying to reach the landmark with highest reward value present in the scene so as to reach the target. A detailed description of an experiment is reported in the following.

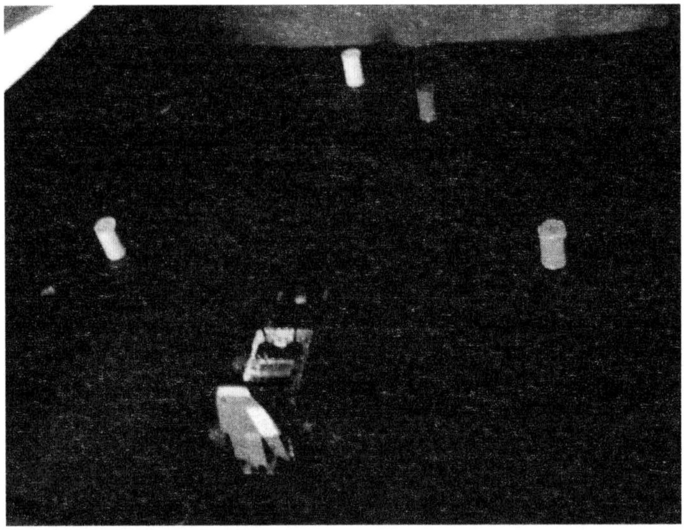

Figure 7. The experiments with the robot are carried out in an arena with cylinder-shape objects of different colors. The black panels are used to increase the color contrast.

The environment for the experiment has been realized as shown in Figure 7. The elements introduced are identical cylinders which may be distinguished through their color.

In the first step the yellow cylinder is chosen among the objects distributed in the field as target. The robot extracts the features necessary to distinguish the target from the other elements included in the field. All the information are stored in the LTM and a new parameter named reward is assigned to this classified object. This parameter is used to identify the importance of the corresponding object within the robot task. In the proposed experiment the robot's aim is to reach the target and then the reward value of the target will be the greatest. A screen shot of the software interface is shown in Figure 8.

The robot is then able to explore the environment analyzing the scenario and storing information about the identified objects. The video, acquired with the webcam, is elaborated on the PC with image processing techniques: segmentation, contour extraction, noise filtering and so on. In this experiment the principal feature is the object color. Another important information acquired through the image processing is the distance between

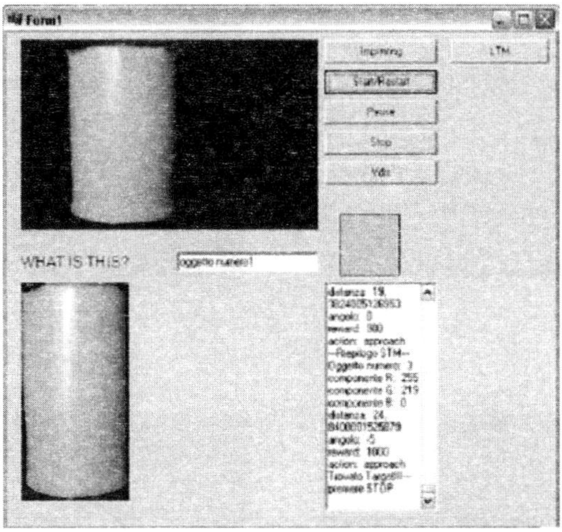

Figure 8. The image elaborations are visualized on the PC. In the considered case the yellow cylinder is classified as target.

the robot and the object: this measure can be carried out since the height of the cylinders is fixed. All this information are allocated in the STM realized with a 'first in first out' (FIFO) buffer. When the imprinted target is localized, the information previously acquired are correlated to it. A Reward value is assigned to a given number of elements stored in the FIFO buffer in relation to the distance to the target. The objects are then stored in the long-term memory as landmarks. The sequence of the last trial is given in Figure 9. The robot, randomly deployed in the field, is able to recognize a landmark, the cyan cylinder, and to reach it. After an exploratory movement, the yellow and red cylinders are identified: the robot examines the Reward value of the two elements and recognizes the first one as a target and then it goes toward it.

Bibliography

R. C. Arkin. *Behavior-Based Robotics.* MIT Press: Cambridge, MA, 1998.

H. Cruse. The evolution of cognition a hypothesis. *Cog. Science*, 27, 2003.

J.J. Gibson. *The Senses Considered as Perceptual Systems.* Houghton Mifflin, Boston, 1966a.

J.J. Gibson. *An Ecological Approach to Visual Perception.* Houghton Mifflin, Boston, 1979b.

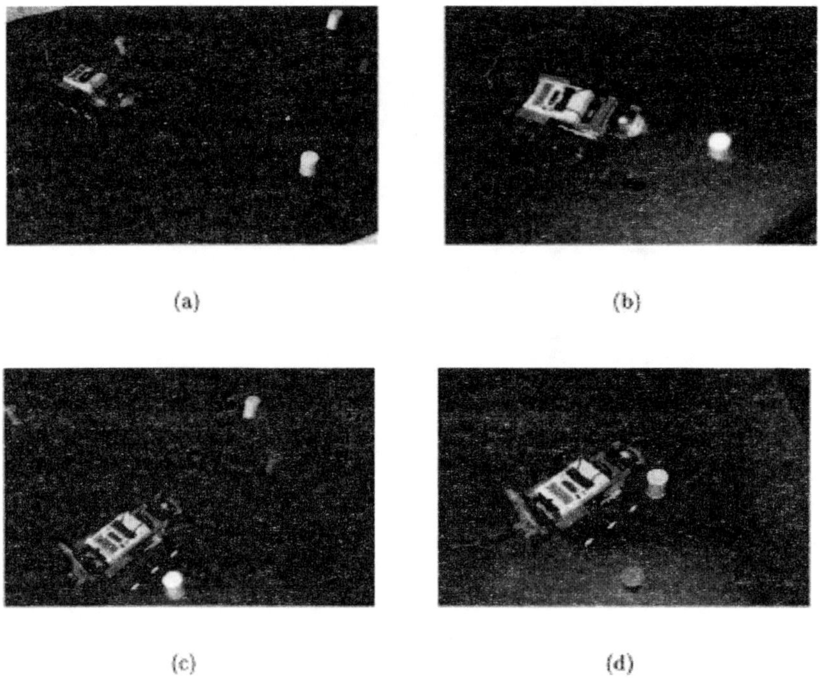

Figure 9. The image sequence describes the path covered by the robot after a learning phase. (a) Random deployment, (b) approach toward a landmark (i.e. cyan cylinder), (c) field exploration and target identification, (d) the robot reaches the target (i.e. yellow cylinder).

R.L. Gregory. Perceptions as hypotheses. *Philosophical Transactions of the Royal Society of London*, Series B, 290:181–197, 1980.

D.O. Hebb. *The Organisation of Behaviour.* Wiley, New York, 1949.

U. Neisser. *Cognition and Reality.* W.H. Freeman, San Francisco, 1976.

I. Pavlov. *Conditioned Reflexes: An Investigation of the Physiological Activity of the Cerebral Cortex.* Oxford University Press, London, 1927.

O.G. Selfridge and J.A. Franklin. The perceiving robot: What does it see? what does it do? In *5th IEEE International Symposium on Intelligent Control*, Germany, 1990.

G. M. Shepherd. *Neurobiology.* Oxford Univ. Press, 1997.

P.F.M.J. Verschure and P. Althaus. A real-world rational agent: Unifying old and new AI. *Cognitive Science*, 27:561–590, 2003.

P.F.M.J. Verschure, B. Krose, and R. Pfeifer. Distributed adaptive control: the self-organization of structured behavior. *Robotics and Autonomous Systems*, 9:181–196, 1992.

P.F.M.J. Verschure, T. Voegtlin, and R.J. Douglas. Environmentally mediated synergy between perception and behaviour in mobile robots. *Nature*, 425:620–624, 2003.

Part III

Practical Issues

Practical Issues of "Dynamical Systems, Wave based Computation and Neuro-Inspired Robots" – Introduction

Adriano Basile and Mattia Frasca

Dipartimento di Ingegneria Elettrica Elettronica e dei Sistemi,
Universitá degli Studi di Catania, Catania, Italy

Abstract This Chapter introduces the topics covered during the practice hours of the course "Dynamical Systems, Wave based Computation and Neuro-Inspired Robots". These practice hours were divided into two parts. Firstly, the course participants were asked to learn the basics on CNNs by using a CNN simulator. Then, they were divided in several groups and a project was assigned to each group. In this Chapter the project objectives are introduced, while in the following Chapters the projects are detailed in the contributions given by the course participants.

1 Introduction

This Chapter introduces the topics discussed during the lab hours of the school "Dynamical systems, wave based computation and neuro-inspired robots" held in Udine, September 22-26, 2003. The practice hours were organized in the afternoon so that they can cover the key points of the topics introduced during the morning lectures. The main objectives of the practice hours were to allow the course participants to familiarize with Cellular Neural Network (CNN) for low and high level control of neuro-inspired robots and to face practical aspects arising in the implementation and control of a neuro-inspired robot. To this aim, the course participants were firstly asked to learn the use of a CNN simulator and then they were divided into groups, and to each group a different project was assigned. These projects will be detailed in the following Chapter, where the contributions by the course participants are given.

In particular, the first part of the practice hours in the school was dedicated to the study of CNN. CNNs (Chua and Yang, 1988b,a; Chua and Roska, 1993; Manganaro et al., 1999) constitute a paradigm for investigating nonlinear phenomena: they provide a framework for the study of wave based computation and a powerful tool to implement the dynamical systems involved in the control of neuro-inspired robots.

After a short presentation on the issues related to the implementation of a CNN simulator, the course participants were asked to implement their own CNN simulator in MATLAB. Then the unofficial MATLAB CNN toolbox, developed at the University of Catania, was introduced. This provides an easy-to-use simulator with programmable instructions. Nonlinear phenomena in CNN, in particular autowave generation and propagation and Turing patterns, were investigated by using this simulator. This first part of the practice hours was intended to provide to the students the background for low and high level control of neuro-inspired robots.

In the second part of the practice hours, the algorithms developed in part I were applied to real robots. Since the amount of available time was limited, simple robots built with LEGO MindStorms™were taken into account. Both wheeled and walking robots were considered. Wheeled robots were used to test the high level control algorithm developed in the first phase. In particular robots driven by wave based computing were discussed. Walking robots controlled by simplified CPG were introduced to experiment the topics covered in the low level locomotion control of neuro-inspired robots. Moreover, the hexapod robots developed at the University of Catania (Arena et al., 2002) were presented.

The choice of using LEGO MindStorms as tools for neuro-inspired robotics was motivated by the following considerations. In robotics the use of low-cost, easy-to-build and re-configurable mobile robot kits allows to achieve educational and research objectives. This approach is gaining interest in the scientific community as witnessed by recent papers on robotics journals (Klassner and Anderson, 2003; Greenwald and Kopena, 2003). They allows to create mobile robot labs (Greenwald and Kopena, 2003) for the study of topics related to mobile robot control. They provide a very efficient and low-cost tool allowing the student to face all the topics connected to the development of a robot: construction (mechanics), control (electronics and computer science) and applications.

In particular LEGO MindStorms offer low-cost, flexibility (they support a lot of sensors, effectors, building blocks), student interest and professional curiosity (Klassner and Anderson, 2003). Thanks to their characteristics building a LEGO robot is not a time-consuming process and can be easily addressed in several practice hours of a 5-day school, allowing the remaining time to focus on the development of the control strategies. For these reasons they are here used to build both wheeled and legged robot to test neuro-inspired control methodologies.

More in details the practice hours were scheduled according to Table 1.

Table 1. Organization of the labs.

Day	Practice	Topics
Monday	Practice 1	Brief introduction on MATLAB
Monday	Practice 2	Implementing a CNN simulator
Monday	Practice 3	Implementing a CNN simulator
Monday	Practice 4	Demo: a simple CNN simulator
Tuesday	Practice 1	Introduction on the CNN toolbox
Tuesday	Practice 2	Using the CNN toolbox for image processing
Tuesday	Practice 3	Using the CNN toolbox for autowaves, Turing patterns
Wednesday	Practice 1	Locomotion control and introduction on projects
Wednesday	Practice 2	Construction of the robot
Wednesday	Practice 3	Construction of the robot
Wednesday	Practice 4	Construction of the robot and set-up of the control framework
Thursday	Practice 1	Control of the robot
Thursday	Practice 2	Control of the robot
Thursday	Practice 3	Control of the robot

The remain of the paper is organized as follows: in Section 2 the practice hours dealing with CNN are discussed; in Section 3 a brief overview on the projects is presented.

2 CNN simulations

2.1 Implementation of a CNN simulator

The first part of the laboratory hours was devoted the implementation of a CNN simulator in MATLAB. After a brief introduction on the main procedures for a CNN simulator, two hours were given to the participants to complete their own simulator. More in detail the participants were asked to complete the following steps:

- Write a MATLAB program to simulate a 1-layer CNN.
- Use the simulator to perform an image processing task: e.g. edge detection on image of Figure 1. The test image is "Quadrati_Ruotati". The templates are the following:

$$A = \begin{bmatrix} 0 & 0 & 0 \\ 0 & 0 & 0 \\ 0 & 0 & 0 \end{bmatrix}; B = \begin{bmatrix} -1 & -1 & -1 \\ -1 & 8 & -1 \\ -1 & -1 & -1 \end{bmatrix}; \quad I = -1$$

- Try other templates in the library (see Roska et al. (1998)).

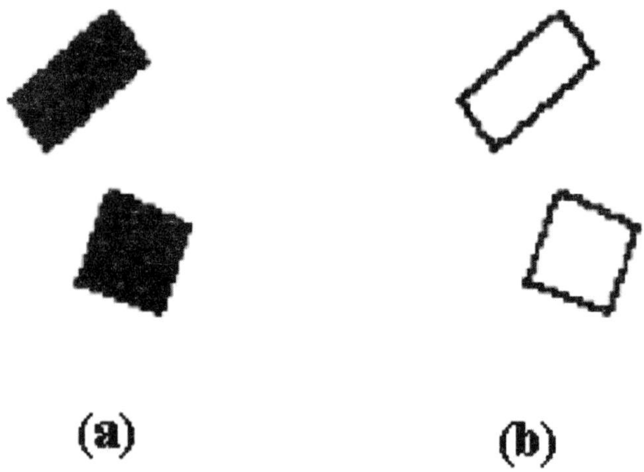

Figure 1. Test image for edge detection.

2.2 DEMO The core of a CNN simulator

The last hour of the first day of practice hours was dedicated to a short demo of the core of a CNN simulator. The CNN simulator written in MATLAB consists of two parts: firstly the object CNN, its templates, initial conditions and neighborhood are defined; then the temporal evolution of the CNN runs.

Definition of the CNN

```
%definition of the cnn

%initial conditions, templates and neighborhood
```

```
cnn.A=[0 0 0 0 1 0 0 0 0]; %3 by 3
cnn.B=[0 0 0 1 1 -1 0 0 0]; %3 by 3
cnn.C=[0 0 0 0 0 0 0 0 0]; %3 by 3
cnn.I=-2;

F=[0 0 0 0 0 0 0 0 0 0 0 0; 0 -1 -1 -1 -1 -1 -1 -1 -1 -1 -1 0; 0
-1 -1 -1 -1 -1 -1 -1 -1 -1 -1 0; 0 -1 -1 -1 -1 1 1 -1 -1 -1 -1 0;
0 -1 -1 -1 1 1 1 1 -1 -1 -1 0; 0 -1 -1 1 1 1 1 1 1 -1 -1 0; 0 -1
-1 1 1 1 1 1 1 -1 -1 0; 0 -1 -1 -1 1 1 1 1 -1 -1 -1 0; 0 -1 -1 -1
-1 1 1 -1 -1 -1 -1 0; 0 -1 -1 -1 -1 -1 -1 -1 -1 -1 -1 0; 0 -1 -1
-1 -1 -1 -1 -1 -1 -1 -1 0; 0 0 0 0 0 0 0 0 0 0 0 0];

% figure; imagesc(F); colormap(gray);

N=12;
cnn.cells=cell(N,N); %N-2 by N-2, + null raws and columns
cnn.X=F; cnn.U=zeros(N); cnn.U=F;
cnn.Y=1/2*(abs(cnn.X+1)-abs(cnn.X-1));

%create the neighborhood for each cell
for i=2:N-1
    for j=2:N-1
        cnn.cells{i,j}.neigh=[i-1 i-1 i-1 i i i i+1 i+1 i+1;
                              j-1 j j+1 j-1 j j+1 j-1 j j+1];
    end %j
end %i
```

Evolution of the cnn

```
%evolution of the cnn
%cycles of the cnn (kij)

deltaT=0.1; %%%%%%%%%%%%%%%% integration step size

for k=1:100 %100 cycles

    cnn.Y=1/2*(abs(cnn.X+1)-abs(cnn.X-1));

    for i=2:(N-1)
    for j=2:(N-1)
        vettoreaggA=[];
        vettoreaggB=[];
        vettoreaggC=[];
        for kk=1:9
```

Table 2. Main commands of the CNN toolbox

Command	Use
`imageprocessing`	Process an image using CNNs
`new_two_layer`	Create a 2-layer CNN struct
`two_layer_processing`	Process a 2-layer CNN (e.g. for Turing patterns)
`new_autowave`	Create a 2-layer CNN struct for autowave simulation
`autowaves_processing`	Process a 2-layer CNN for autowave simulation
`draw_cnn_structure`	Draw the connections between cells of a CNN
`modify_connection`	Create a new connection between cell ij and cell kl

```
            vettoreaggA=[vettoreaggA cnn.Y(cnn.cells{i,j}.neigh(1,kk),
            cnn.cells{i,j}.neigh(2,kk))];
            vettoreaggB=[vettoreaggB cnn.U(cnn.cells{i,j}.neigh(1,kk),
            cnn.cells{i,j}.neigh(2,kk))];
            vettoreaggC=[vettoreaggC cnn.X(cnn.cells{i,j}.neigh(1,kk),
            cnn.cells{i,j}.neigh(2,kk))];
        end %kk
        xdot=-cnn.X(i,j)+cnn.A*vettoreaggA'+cnn.B*vettoreaggB'
            +cnn.C*vettoreaggC'+cnn.I;
        cnn.X(i,j)=cnn.X(i,j)+deltaT*xdot;
    end %j
    end %i
end %k
```

2.3 The CNN toolbox

The second day of practice hours was dedicated to simulating more complex CNN structures. To this aim the CNN toolbox was introduced.

The unofficial MATLAB CNN toolbox has been developed at the University of Catania. It is intended to provide basic MATLAB functions and a Graphic User Interface for simulating CNN in MATLAB. The most important commands are given in Table 2. The help for these functions is called from the MATLAB prompt, as usually, by typing: `help` *name_function*.

Three different structures are defined: a `cnn` struct for image processing (this is a 1-layer CNN, to access for example to the A template it is possible to use `cnn.A`); a `cnn` struct for simulation of autowaves (this a 2-layer CNN, but the structure is optimized for autowave simulation, in fact all the coefficients of templates A_{12} and A_{21} are zero except the central coefficients); a `cnn` struct for 2-layer simulation (this is a general 2-layer CNN, it can be used for example to simulate Turing Patterns).

The target of using the CNN toolbox is understanding CNN for image processing and for the study of nonlinear phenomena. The practice hours were organized by considering the following steps:

- IMAGE PROCESSING: Write a simple program based on the CNN toolbox for the complex image processing task of Figure 2 (Chua and Roska, 2002).
- Simulation of autowave propagation: Try several initial conditions, several parameters, ...
- TURING PATTERNS: Simulate Turing Patterns arising from different initial conditions, different parameter values, ...

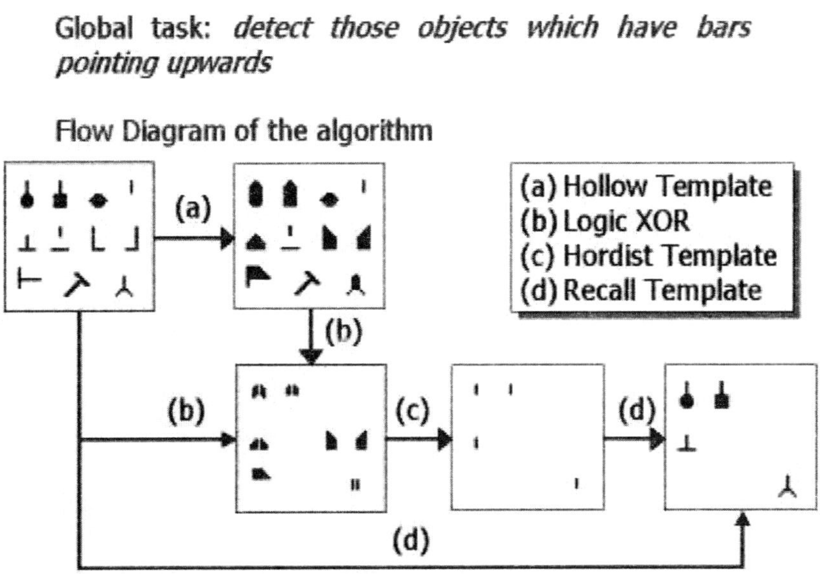

Figure 2. Example of a CNN sequence of template for a complex image processing task.

Development of the image processing task The following MATLAB program can be used to detect objects pointing upwards:

```
clear all
```

```
a_hollow =[  0.5      -0.5     0.5
             0.5        2      0.5
             0.5       0.5     0.5];

b_hollow =[ 0  0  0
            0  2  0
            0  0  0];

c_hollow =[ 0      0      0
            0     -1      0
            0      0      0];
bias_hollow =3;

a_hordist=[  0      0     0  0  0
             0      0     0  0  0
             0      0     2  0  0
             0      0     0  0  0
             0      0     0  0  0];

b_hordist=[ 0      0       0  0  0
            0      0       0  0  0
            0.25  -0.25    0  0  0
            0      0       0  0  0
            0      0       0  0  0];

c_hordist=[  0      0      0  0  0
             0      0      0  0  0
             0      0     -1  0  0
             0      0      0  0  0
             0      0      0  0  0];
bias_hordist =-1.5;

a_recall =[  0.5      0.5     0.5
             0.5       4      0.5
             0.5      0.5     0.5];

b_recall =[ 0  0  0
            0  4  0
            0  0  0];

c_recall =[ 0      0      0
            0     -1      0
            0      0      0];

bias_recall =2.1;
```

```
input=im2matrix;

%hollow
figure(1); imagesc(-input); colormap(gray); title('input');
state=input; figure(2); title('hollow result')
[cnn]=imageprocessing(state,input,a_hollow,b_hollow,
c_hollow,bias_hollow,.1,200,2,10,0) %hollow

%xor
xorresult=xor((-input+1)/2,cnn.cnn_output_image)*2-1; figure(3);
imagesc(-xorresult); colormap(gray); title('xor result');

%hordist
figure(4); title('hordist result')
[cnn]=imageprocessing(xorresult,xorresult,a_hordist,b_hordist,
c_hordist,bias_hordist,.1,200,2,10,0) %hordist

%recall
figure(5); title('recall result') image1=cnn.cnn_output_image*2-1;
[cnn]=imageprocessing(-image1,input,a_recall,b_recall,
c_recall,bias_recall,.05,200,2,10,0) %recall
```

3 Brief overview of the projects

In the last two days of practice hours the participants were divided into several groups and developed neuro-inspired projects. Several projects make use of simple LEGO robots to implement neuro-inspired algorithms. Therefore a short introduction on the control of LEGO robots was provided to all the participants.

3.1 Introduction on the control of LEGO robots

There are two ways in which LEGO Mindstorms can be controlled: autonomous way and direct control.

In the first case the robot has its own program downloaded in the microcontroller. In this case the robot is autonomous, but the control software is limited by the memory of the RCX. We suggest to use NQC code for this purpose.

As regards the direct control (in which the LEGO is directly controlled by the PC) many softwares have been developed. A very efficient way is to use C or C++ and include libraries from LEGO Mindstorms in the developed

Table 3. LegoConsole Program.

Command	**Use**
/ECHO:ON	Enable the Echo on Screen (default)
/ECHO:OFF	Disable the Echo on Screen
G	Go on
TL	Turn Left
TR	Turn Right
B	go Back
TIME	instruction time

Table 4. LegoSens Program.

Command	**Meaning**
/ECHO:D	switch the echo on Display (default)
/ECHO:F	switch the echo on File (to be conbined with /F option)
/F:filename	specify the file name
RLSx	Read Light Sensor number **x**
RTSx	Read Touch Sensor number **x**
RRSx	Read Rotation Sensor number **x**
	x must be a valid number from 1 to 3

project. Here we propose a higher level approach. Since the algorithms from previous labs have been developed in MATLAB, we propose to use from the MATLAB environment an .exe file implementing simple routines for the LEGO robot. This is called LegoConsole.exe. It controls the robot so that it moves in one of the direction specified by the parameters of Table 4 for a given amount of time.

These commands can be used from MATLAB prompt (or functions) as follows:

```
!LegoConsole [/ECHO:[ON|OFF]] COMMAND TIME
```

Analogously to read the inputs of the RCX microprocessor LegoSens.exe can be used.

The syntax of this command is the following:

```
!LegoSens [/ECHO:[D|F] /F:filename] COMMAND
```

3.2 Neuro-inspired projects

The following projects have been developed during the school:

- "Locomotion control of a hexapod by Turing patterns" by M. Pavone, M. Stich, B. Streibl;
- "Visual Control of a Roving Robot based on Turing Patterns" by P. Brunetto, A. Buscarino, A. Latteri;
- "Wave-based control of a bio-inspired hexapod robot" by F. Danieli, D. Melita;
- "Cricket phonotaxis: simple Lego implementation" by P. Crucitti and G. Ganci;
- "CNN-based control of a robot inspired to snakeboard locomotion" by G. Aprile, M. Porez, M. Wrabel.
- "Cooperative behavior of robots controlled by CNN autowaves" by P. Crucitti, G. Dimartino, M. Pavone, C. Presti.

The first two projects deal with the implementation of the control strategy based on Turing Patterns (Arena et al., 2003). The approach is based on a CNN generating Turing patterns, these patterns depend on the CNN initial conditions that are related to the information coming from sensors. In the first case this approach is applied to a roving robot using a vision camera. In the second case Turing patterns are used to control a legged robot.

Another project is related to the implementation of cricket phonotaxis on a LEGO robot. The behavior of the neuro-inspired robot is controlled according to the sound localization behavior shown by crickets.

Two other projects deal with the implementation of the control strategy illustrated in Adamatzky et al. (2004). In this case the environment in which the robot moves is mapped into a CNN in which autowaves propagate. Obstacles and target are the autowave sources, and autowaves controls the robot navigation. The first project shows how this strategy can be used to control a team of cooperating roving robots, the second one illustrates the implementation of the methodology on a legged robot.

Another project deals with the construction and the control through a CNN-based Central Pattern Generator of a robot with snake-like locomotion.

The details of the projects are described in the following Chapters.

Bibliography

A. Adamatzky, P. Arena, A. Basile, R. Carmona-Galan, B. De Lacy Costello, L. Fortuna, M. Frasca, and A. Rodriguez-Vazquez. Reaction-diffusion navigation robot control: from chemical to vlsi analogic processors. *IEEE Trans. on Circuits and Systems I*, 51 (5): 926–939, May 2004.

P. Arena, L. Fortuna, and M. Frasca. Attitude control in walking hexapod robots: An analogic spatio-temporal approach. *Int. J. Circ. Theor. Appl.*, 30:349–362, 2002.

P. Arena, A. Basile, L. Fortuna, M. Frasca, and L. Patané. Implementation of turing patterns for bio-inspired motion control. In *Proc. of IEEE Int. Conference ISCAS03*, volume 3, pages 842–845, 2003.

L. O. Chua and T. Roska. The CNN paradigm. *IEEE Transactions on Circuits and Systems*, 40:147–156, 1993.

L. O. Chua and L. Yang. Cellular Neural Networks: Applications. *IEEE Trans. on Circuits and Systems I*, 35:1273–1290, October 1988a.

L. O. Chua and L. Yang. Cellular Neural Networks: Theory. *IEEE Trans. on Circuits and Systems I*, 35:1257–1272, October 1988b.

L.O. Chua and T. Roska. *Cellular Neural Networks & Visual Computing.* University Press, 2002.

L. Greenwald and J. Kopena. Mobile robot labs. *IEEE Robotics and Automation Magazine*, 10 (2):25–32, 2003.

F. Klassner and S. D. Anderson. Lego mindstorms: Not just for k-12 anymore. *IEEE Robotics and Automation Magazine*, 10 (2):12–18, 2003.

G. Manganaro, P. Arena, and L. Fortuna. *Cellular Neural Networks: Chaos, Complexity and VLSI processing.* Springer-Verlag, 1999.

T. Roska, L. Kek, L. Nemes, A. Zarandy, M. Brendel, and P. (Eds.) Szolgay. *CNN Software Library (Template and Algorithms).* Version 7.2, 1998.

Locomotion control of a hexapod by Turing patterns

Marco Pavone, * Michael Stich, † and Bernhard Streibl ‡

* Scuola Superiore di Catania, Università degli Studi di Catania, Via San Paolo 73, 95123 Catania (Italy).

† Instituto Pluridisciplinar, Universidad Complutense de Madrid, P. Juan XXIII 1, 28040 Madrid (Spain).

‡ Institut für Strömungslehre und Wärmeübertragung, Technische Universität Wien, Resselgasse 3, 1040 Wien (Austria).

Abstract In this paper, the reflexive behavior of a biomorphic adaptive robot is analyzed. The motion generation of the robot is governed by a Reaction-Diffusion Cellular Neural Network (RD-CNN) that evolves towards a Turing pattern representing the action pattern of the robot. The initial conditions of this RD-CNN are given by the sensor input. The proposed approach is particularly valuable when the number of sensors is high, being able to perform data compression in real-time through analog parallel processing. An experiment using a small 6-legged robot realized in Lego MindStorms™ with three sensors is presented to validate the approach. A simulated 3x3 CNN is used to control this hexapod.

1 Introduction

In current robotics it is a major enterprise to create robots capable of moving autonomously in an unknown environment. One major obstacle in the design of such robots is the large number of sensors that may require a long time to process the gathered information. One inspiration to overcome this problem arises from the biological paradigm of the Central Pattern Generator (CPG), the part of the nervous system responsible for locomotion control in animals. From a behavioral point of view, the CPG can be modelled as a hierarchical structure of several neuron groups: The Command Neurons (CNs), due to input from the sensory or the central nervous system (CNS), activate the Local Motion Generating Neurons (LMGNs) that generate the appropriate signals for the type of locomotion induced by the CNs (Wilson, 1972; Stein, 1978; Calabrese, 1995). The hierarchical structure of the CPG is schematically shown in Figure 1.

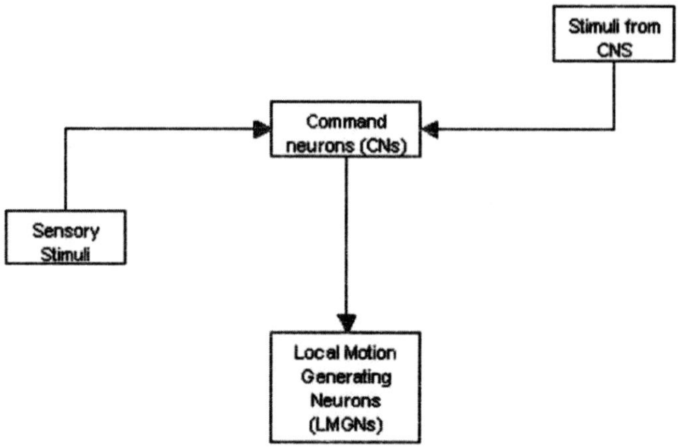

Figure 1. Scheme of the CPG.

Both LMGNs and CNs can be modelled by Reaction-Diffusion Cellular Neural Networks (RD-CNNs) (Chua, 1995). On one hand, such networks can produce particular waves, so-called autowaves, able to drive locomotion effectors. On the other hand, RD-CNNs also allow the presence of Turing patterns, which are stationary and spatially-periodic patterns. The main idea of this paper is that locomotion schemes generation (corresponding to CNs) can be realized by using a CNN structure able to generate Turing patterns: each steady state pattern realizes a particular locomotion scheme (Arena et al., 2003).

In order to realize different locomotion types, the CNN grid implementation of the CNs must possess a discrete quantity of steady state configurations. Furthermore, the particular steady state condition corresponding to a particular locomotion type must be directly controlled in parallel by different types of sensory inputs.

For the bio-inspired robot described in this paper, the locomotion control by Turing patterns is simulated on a PC, while the other aspects are treated as in traditional robotics.

2 The robot structure

To validate the approach stated above, we have realized an hexapod in Lego MindStorms^{TM}. This simple and flexible system is well-suited for a fast implementation, avoiding many mechanical problems arising in the design of a robot. The bio-inspired robot is composed by two parts: the front and the back part. The structure of the robot is schematically shown in Figure 2. Four legs are connected to the back part and two legs are connected to the front part. Change of the direction of motion (rotation) is achieved by turning the front part with respect to the back part. On the back part, the RCX microprocessor and two motors are placed. The former is used for forward movement (motor C), the latter for rotation (motor B). On the front part, one motor for forward motion (motor A) and three light sensors able to detect obstacles are placed. The sensory input is discussed in the next section. All the computations are performed on a PC; the data exchange between the microprocessor and the PC is realized via an infrared communication. A photo of the robot is shown in Figure 3. Motor synchronization is adjusted in such a way that the robot performs fast gait motion, as shown in Figure 4.

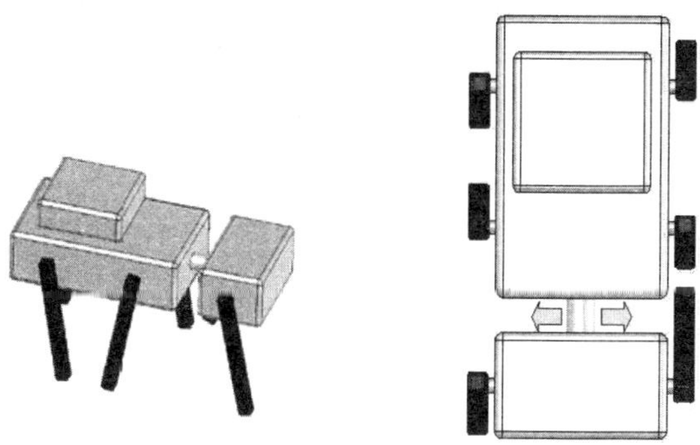

Figure 2. Robot scheme.

3 CNN patterns and action patterns

We have used a small RD-CNN consisting of 3x3 cells. Although the variables can take continuous values, there are only eight steady state (Turing) patterns for such a network, which are easily distinguishable (Arena et al., 1998). For each initial condition, represented by the obstacle configuration as sensed by the robot, the CNN evolves towards a Turing pattern. The mapping between sensors and CNN reflects the spatial arrangement, as shown in Figure 5.

Once the system has reached the final Turing pattern, the processor

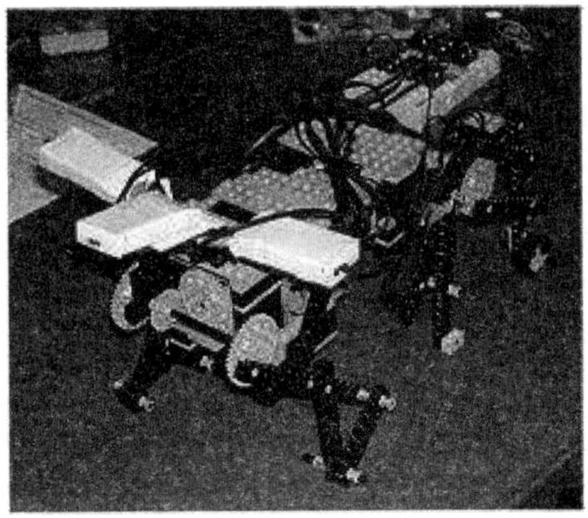

Figure 3. Photo of the robot.

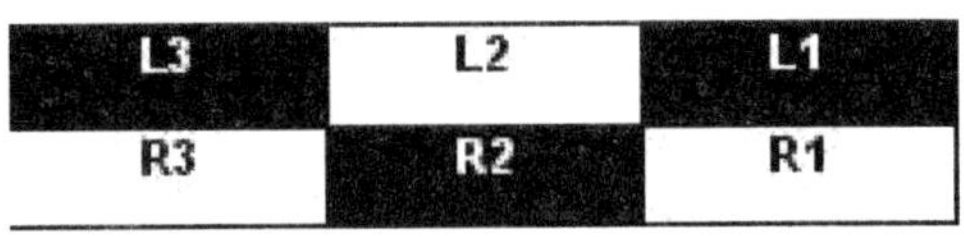

Figure 4. Gait scheme. L1 (R1) controls the first left (right) leg, and so on. Legs which are in the same phase of motion are denoted by the same color.

decides upon the locomotion scheme, i.e., each Turing pattern encodes a specific locomotion scheme Arena et al. (1999). For the robot discussed in this paper, the encoding is explicitly given, as reported in Figure 6, where triangle and circles stand for robot and obstacles respectively. The locomotion is easily implemented on the RCX microprocessor by the sequence of three motor actions. In Figure 7, these actions are described without loss of generality for the task “go east”.

4 Discussion

The robot presented in this paper is able to perform locomotion based on CPG paradigm, which offers clear advantages compared to the traditional robotics approach (see also other papers of this volume). We were able to demonstrate that this bio-inspired robot can process sensory data very efficiently, achieving parallel sensory data compression and real-time action decision. The implementation has been done by using a RD-CNN mimicking

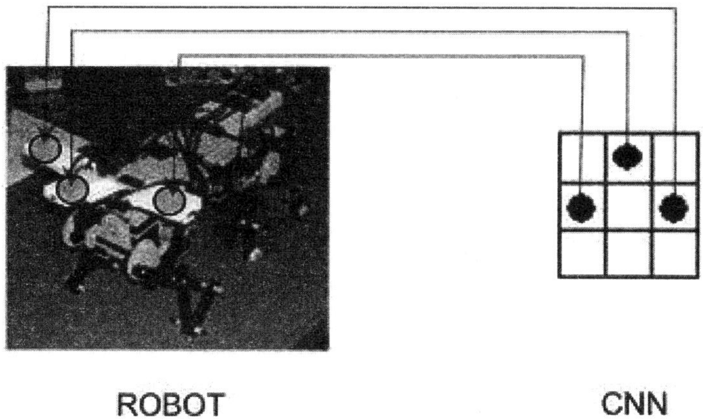

Figure 5. Mapping of the sensors to the CNN.

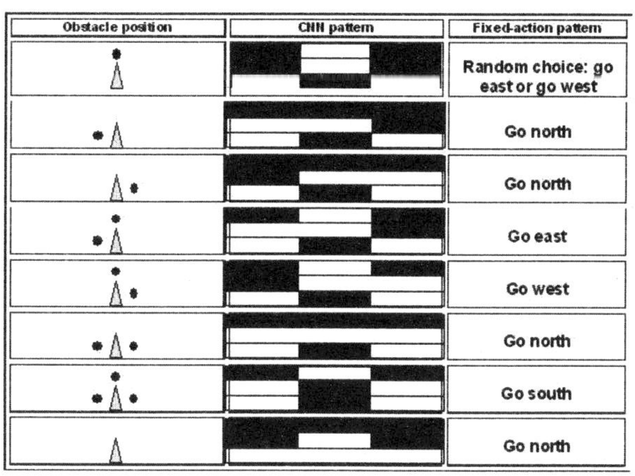

Figure 6. Encoding actions by Turing patterns.

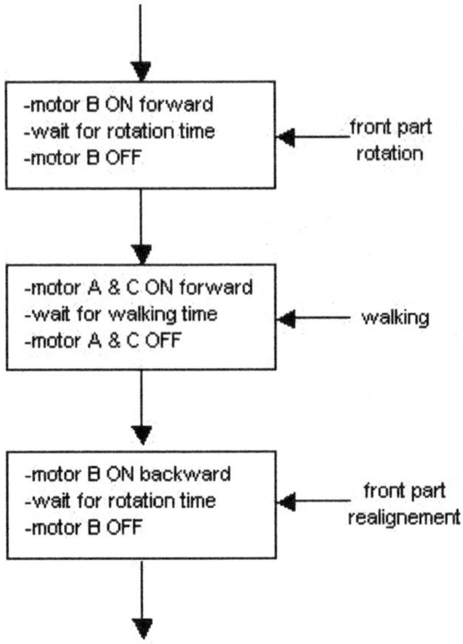

Figure 7. Flow chart of the task "go east".

the behavior of Command Neurons of a CPG. Clearly, for heterogeneous sensory input, data fusion can be realized in a similar way.

Furthermore, such an approach is robust with respect to the ubiquitous presence of noise in real applications. This robustness refers to both, noisy input information, e.g. due to thermal fluctuations, and unavoidable changes of the components' properties. For the robot described above, the assignment of action for a given Turing pattern has been considered given a priori. This implementation has been chosen for convenience. It is possible to use other schemes for the same task, such as artificial neural networks, allowing for the incorporation of learning schemes, e.g. through Kohonen maps.

Bibliography

P. Arena, M. Branciforte, and L. Fortuna. A cnn based experimental frame for patterns and autowaves. *Int. J. on Circuit Theory and Appls.*, 26: 635–650, 1998.

P. Arena, M. Branciforte, and L. Fortuna. Reaction-diffusion cnn algorithms to generate and control artificial locomotion. *IEEE Trans. on Circuits and Systems, Part I*, 46:259–266, 1999.

P. Arena, A. Basile, L. Fortuna, M. Frasca, and L. Patané. Implementation of turing patterns for bio-inspired motion control. In *Proc. of IEEE Int. Conference ISCAS03*, volume 3, pages 842–845, 2003.

R. L. Calabrese. Oscillation in motor pattern-generating networks. *Curr. Opin. Neurobiol.*, 5:816–823, 1995.

L. O. Chua. Special issue on nonlinear waves, patterns and spatio-temporal chaos. *IEEE Trans. on Circuits and Systems, Part I*, 42(10), 1995.

P. S. G. Stein. Motor systems, with specific reference to the control of locomotion. *Ann. Rev. Neurosci.*, 1:61–81, 1978.

D. M. Wilson. Genetic and sensory machanisms for locomotion and orientation in animals. *Am. Sci.*, 60:358–365, 1972.

Visual Control of a Roving Robot based on Turing Patterns

Paola Brunetto,Arturo Buscarino and Alberta Latteri
Dipartimento di Ingegneria Elettrica Elettronica e dei Sistemi,
Universitá degli Studi di Catania, Catania, Italy

1 Introduction

Turing Patterns (TP) could be useful in robot control. In particular, using a webcam on board the robot, a snapshot of the environment could be set as initial condition of a CNN that will generate TP. Analysing the TP aroused is possible to control the robot avoiding obstacles.

TPs are used as a link between the real environment and the perceived environment. In biological systems (such as animals) the processing of a sensorial information starts directly while sensing it: for example, the first processing of a viewed image is made up immediately by retina. Due to this a really biologically inspired control based on vision should make use of a parallel hardware acting directly on the perceived environment image. The approach based on TPs can overcame this problem because TPs are generated by a nonlinear parallel analog system such Cellular Non-linear Network (CNN).

The approach introduced in this work starts from the methodology discussed in Arena et al. (2003). In that paper TPs are used to control the behavior of a reactive robot avoiding obstacles on the basis of a series of sensors. A small CNN (3x3) is used. The initial conditions for the TP generation are set using a 3x3 b/w image built on the information of three light sensors located on the frontal and lateral side of the robot. The generated TP represents a fixed action for the robot to avoid obstacles. In this case since the CNN is 3x3, the number of possible TPs is relatively small.

In this paper the sensorial stimulus comes from the on-board camera. Using a conceptually different kind of input requires a generalization of the approach. First of all, since the size of the CNN is now much bigger than in the case reported in Arena et al. (2003), the number of possible TPs is greater. This requires a generalization of the methodology to fix the correspondence between each pattern and a given action for the robot.

2 Mechanical Structure

The robot is built using LEGO® MindStorms (Klassner and Anderson, 2003). It has six wheels actuated by 2 motors: so the robot is able to turn. On the machine a simple LEGO® WebCam connected to the control PC through the USB port is installed. The robot movements are controlled in the direct control way (i.e. by using the infrared device connected to the PC). The software used to communicate with the robot is the LEGO Console.

3 Control Algorithm

TPs are generated starting from an initial condition that represents a real snapshot of the environment: the starting image is taken by a webcam and then processed by a CNN generating a large amount of possible TPs, so it is impossible to completely classify the allowed TPs and, consequently, the corresponding actions. The big advantage is that now the control algorithm has a real perception of the surrounding area and is able to surely recognize the presence of an obstacle. The robot control is simulated using the Matlab CNN toolbox.

The webcam takes a snapshot of the environment with possible obstacles. The image is resized to a 23x28 one and converted in b/w: in this way the obstacle is in black while the free space is in white. Generally light effects on the body of the obstacle produce white spots on the black figure, to avoid this the first image processing operation is a CNN Hollow Template that fills the white holes.

The whole control algorithm is shown in Figure 1 and is based on the detection of horizontal lines in the TP. To do this the size of the CNN showing TPs is different from that of the CNN elaborating the perceived image. In fact as initial condition of the CNN showing TPs an image obtained by adding five horizontal white lines on the top of the webcam image. This assures the presence of horizontal black lines even in the case of a pattern generated by an obstacle filling the whole image height. This procedure changes the sensorial perception by using a different mapping of the space: the viewed image is traduced to what the robot really senses, similarly to what happens in biological systems.

The 28x28 image obtained is set as initial condition to the TP generator, a 28x28 CNN array. The CNN will evolve producing a stable TP. The presence of an obstacle is thus revealed by horizontal lines in the TP (this because horizontal lines are produced by the horizontal contour of an obstacle) so they are isolated applying the "HorizontalLineDetection" Template.

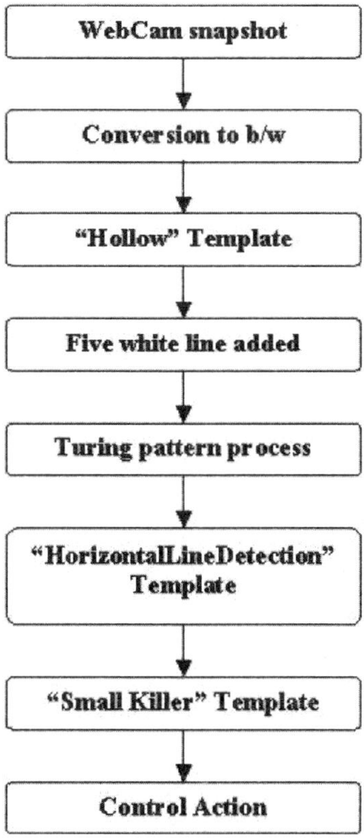

Figure 1. Control algorithm.

The image is also treated with the "SmallKiller" template to erase isolated black spot.

The final image is split in six partitions and the number of black pixel in the three upper parts is calculated. The partition containing the greater number of black pixels indicates the position of the obstacle, so the control action directs the robot to the opposite side. This mathematical clustering of the fixed action, associated to a particular kind of TP, is imposed by the large number of different possible TPs depending on the initial condition image. By using the modified LEGO Console the control command is passed to the robot.

When the original snapshot is completely white, no obstacles are detected and the corresponding TP is white: in this case the robot just proceeds in the same direction without turning. When the obstacle is detected in central position it was chosen to make the robot turn on right side. After each movement the webcam take another snapshot and the algorithm begins again.

Figures 2-8 show experimental results. In particular Figure 2 represents the original gray-scale image taken by the webcam, it is possible to see the obstacle in left position.

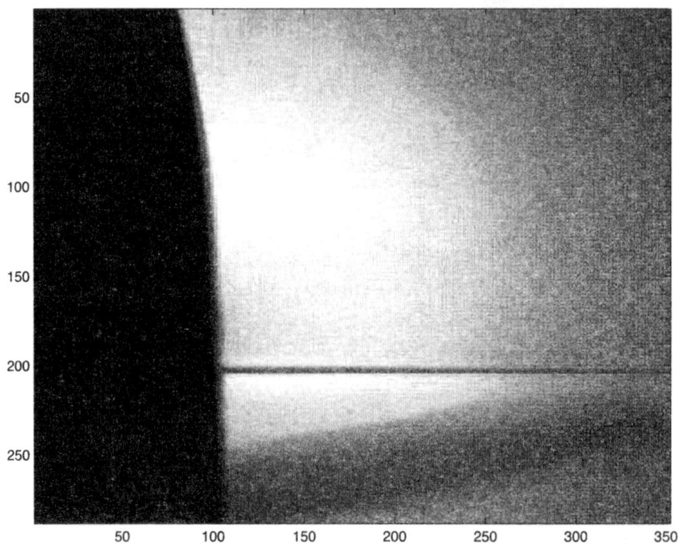

Figure 2. Original image. The obstacle is in left position.

Figure 3 shows the converted b/w image, as it can be noticed there are still imperfections due to light effects. To remove these holes an hollow template is applied, Figure 4 shows the result of this CNN-based image processing: all the white spots are filled.

Figure 5 represents the mapping in the new space in which the TP evolves, with respect to the image shown in Figure 4 it has five white lines added on its top. This picture is set as initial condition of the TPs generator that produce the TP showed in Figure 6.

Finally the last steps of the control algorithm are related to the problem of detecting the horizontal lines. The image representing the final TP is treated with a HorizontalLineDetection CNN template. The result is shown in Figure 7. Figure 8 shows the cleaned final image obtained using SmallKiller CNN treatment.

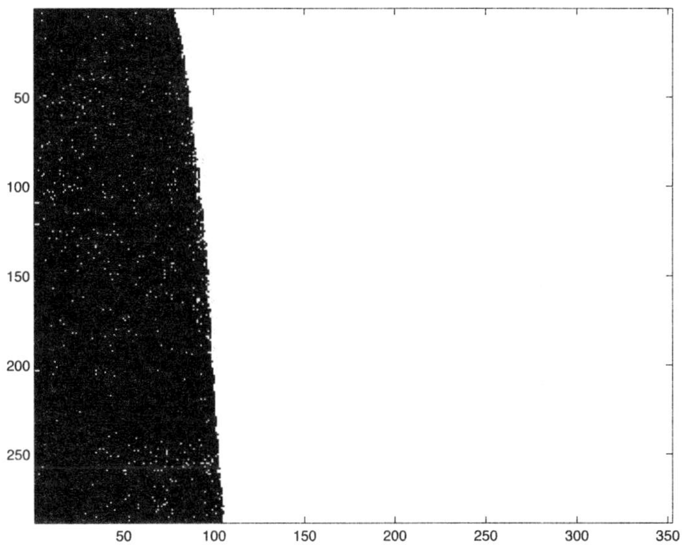

Figure 3. B/w image, some white spots appear inside the obstacle shape.

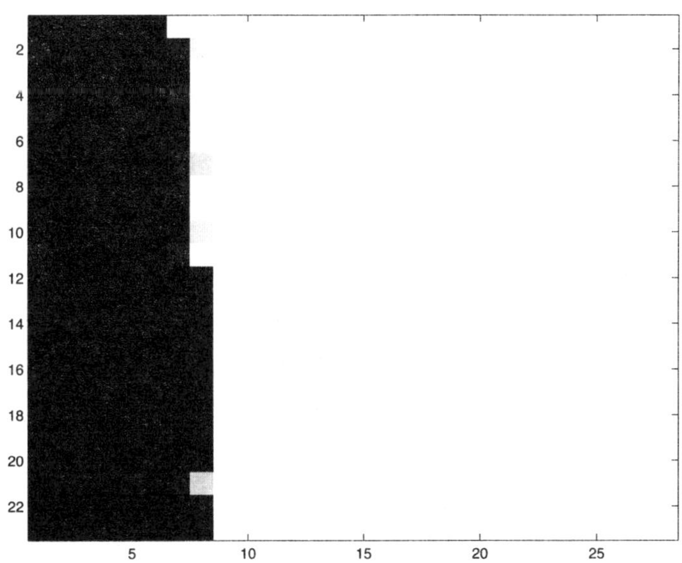

Figure 4. Image after hollow treatment, white spots are filled.

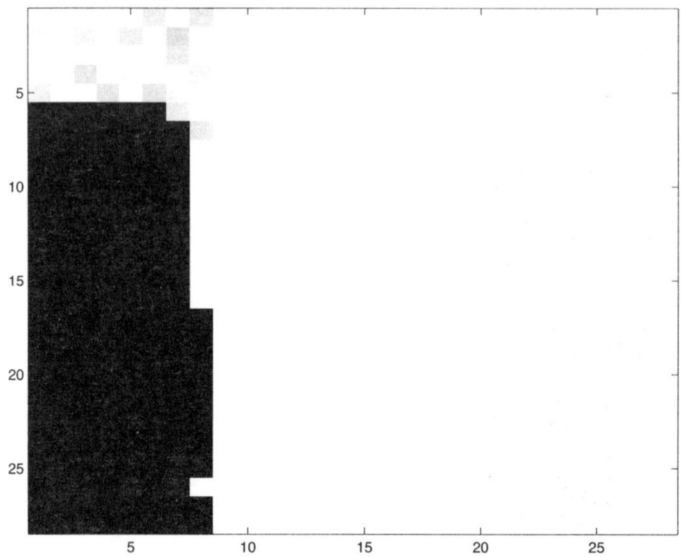

Figure 5. The initial condition for the CNN showing TPs is represented by Fig. 4 with five additional horizontal lines.

Figure 6. The final Turing pattern.

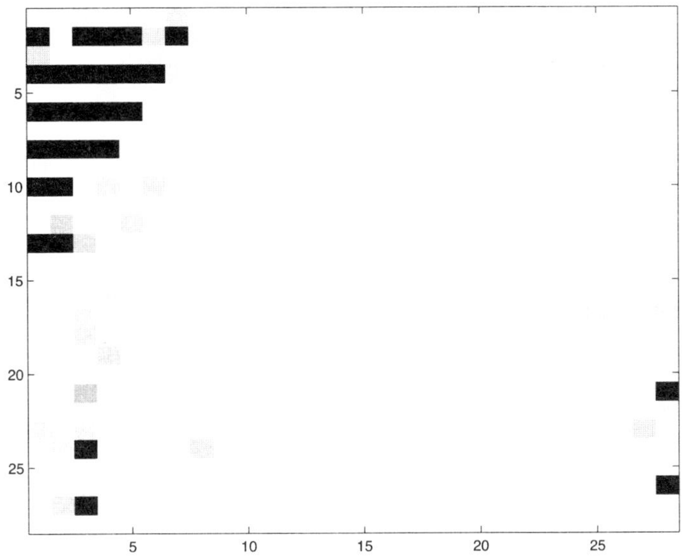

Figure 7. Result obtained applying the Horizontal Line Detection template to Fig. 6.

4 Conclusions

In this paper a simple reactive visual control algorithm based on Turing patterns of a wheeled robot is described. Starting from a snapshot of the surrounding area, the proposed control algorithm has a perception of the environment and is able to make the best decision to avoid the obstacle. This approach satisfies the needs of a real biologically inspired robotics because all the algorithm steps could be make by a non-linear parallel analog system (CNN) that is able to model real biological system behaviour. Following this, it is possible to think an adaptive law to teach the algorithm to recognize the correct action related to a Turing pattern without using the calculation of the number of black pixels in the final image.

Bibliography

P. Arena, A. Basile, L. Fortuna, M. Frasca, and L. Patané. Implementation of turing patterns for bio-inspired motion control. In *Proc. of IEEE Int. Conference ISCAS03*, volume 3, pages 842–845, 2003.

F. Klassner and S. D. Anderson. Lego mindstorms: Not just for k-12 any-

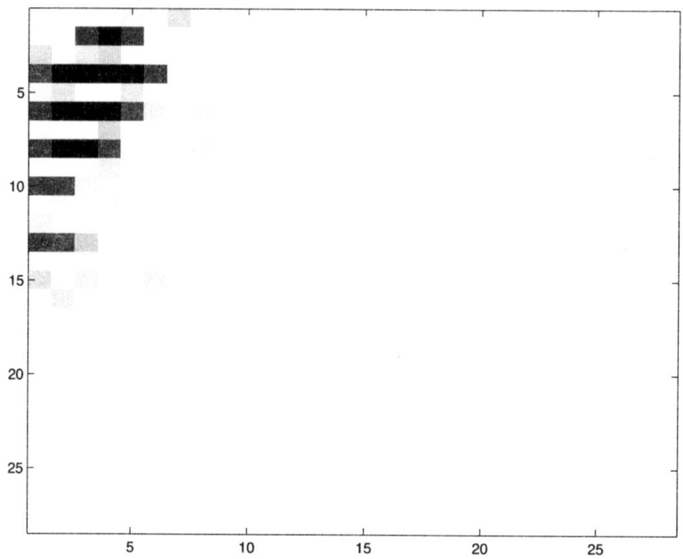

Figure 8. Final image. Black spots in Figure 7 have been deleted by using the Small Killer template.

more. *IEEE Robotics and Automation Magazine*, 10 (2):12–18, 2003.

Wave-based control of a bio-inspired hexapod robot

Fabio Danieli and Donato Melita
Dipartimento di Ingegneria Elettrica Elettronica e dei Sistemi,
Universitá degli Studi di Catania, Catania, Italy

Abstract The main idea of this work is to merge locomotion based on neural approach of Rexabot robot and real-time wave-based navigation in a complex, dynamically changing environment.

1 A framework for the control of a bio-inspired hexapod robot

In this work we used a robotic system that mimics the locomotive behavior of a cockroach insect by using a neural approach. As discussed in Arena et al. (2002) a Cellular Neural Network (CNN) can be used to coordinate the movement of each joint: the dynamics of the system is determined by the number of cells composing the network and by their connections. A framework to control a bio-inspired hexapod robot has been used to reach the purpose of creating a useful prototype to experiment different locomotion patterns and more complex types of control: in fact the dynamics of the network is solved by a computer program called CNNLab that allows, as the main feature, the creation of custom networks. This program is not only able to simulate the behavior of a CNN to generate the proper control signals for the locomotion of the robot but also could take into account the information coming up from the sensors, mounted on the robot, in order to create an adaptive control. The hardware part of this project (schematically depicted in Figure 1) consists of 18 electric servomotors (3 for each leg) climbed on an aluminium structure, an ADXL202 gravity sensor (for the attitude control), 6 optomechanical contact sensors (one for each leg) realized ad hoc for this application and three electronic boards: a board containing the SD20 (a pre-programmed PIC working on the I2C bus able to generate the PWM signals to control each of the servomotors), a board containing an A/D converter (MAX127) also working on the I2C bus and an interface board which connects the I2C bus with the parallel port (LPT) of a normal PC. This framework was made in the attempt of creating a useful environment to implement and test many different kinds of control taking advantage of the flexibility of a computer based approach in order to find new solutions for the bio-inspired robot locomotion problems. A photo of the robot controlled by CNNLab is shown in Figure 2.

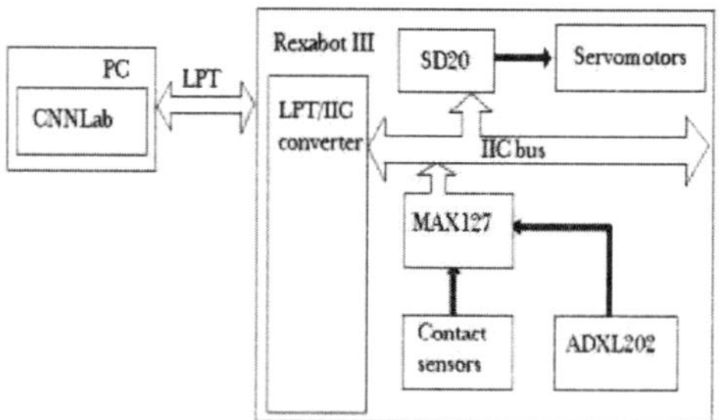

Figure 1. Hardware implementation.

Figure 2. A photo of the hexapod robot controlled by CNNLab.

2 Reaction-Diffusion navigation robot control

In order to realize navigation and localization of Rexabot the well known technique (Adamatzky et al., 2004) based on RD-CNNs was used: mapping robot arena on a CNN, obstacles and targets can be considered as sources of repulsive and attractive autowaves, respectively. A simple AND can detect wavefront collisions with the robot and this information is used for the trajectory generation.

3 Merging software and hardware

The flow diagram of the CNN algorithm is shown in Figure 3. In order to realize this algorithm, a CNN Matlab Toolbox, developed at the University of Catania, was used; it permits to simply elaborate robot arena picture, acquired through a on-board webcam, and to obtain the correct signals to drive Rexabot. An interface between CNN Matlab Toolbox and CNNLab is required to correctly control robot as shown in Figure 4.

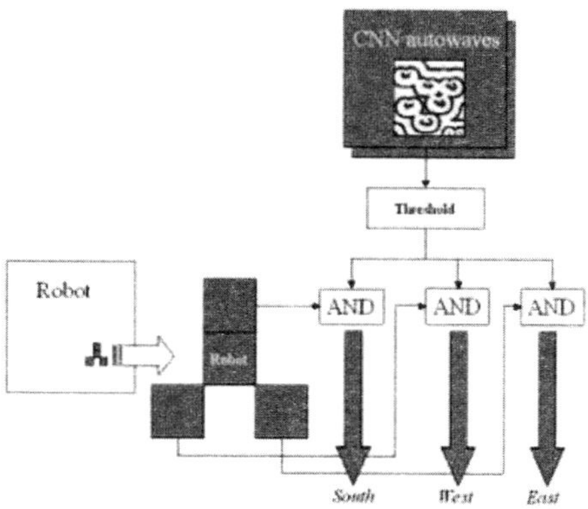

Figure 3. Flow diagram of the algorithm.

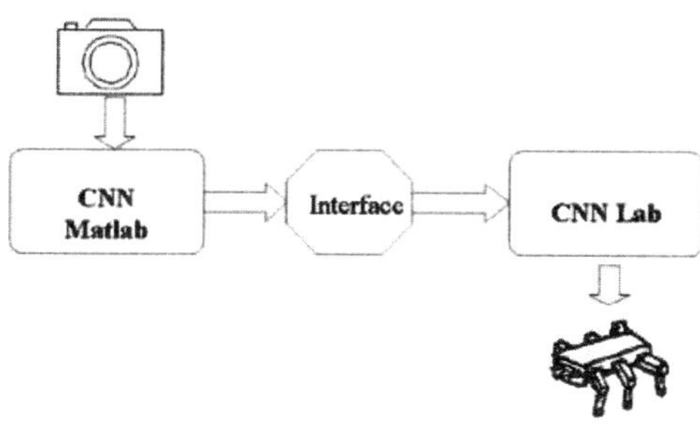

Figure 4. Full implementation diagram.

Bibliography

A. Adamatzky, P. Arena, A. Basile, R. Carmona-Galan, B. De Lacy Costello, L. Fortuna, M. Frasca, and A. Rodriguez-Vazquez. Reaction-diffusion navigation robot control: from chemical to vlsi analogic processors. *IEEE Trans. on Circuits and Systems I*, 51 (5): 926–939, May 2004.

P. Arena, L. Fortuna, and M. Frasca. Attitude control in walking hexapod robots: An analogic spatio-temporal approach. *Int. J. Circ. Theor. Appl.*, 30:349–362, 2002.

Cricket phonotaxis: simple Lego implementation

Paolo Crucitti* and Gaetana Ganci†
* Scuola Superiore di Catania, Catania, Italy
† Dipartimento di Ingegneria Elettrica Elettronica e dei Sistemi, Universitá degli Studi di Catania, Catania, Italy

Abstract Female crickets are able to recognise and localise their mates, using a simple neural structure that implements phonotaxis. In this paper we show simple Lego implementation of this biological model.

1 Introduction

The main idea of bio-inspired robotics is that of drawing inspiration from the efficient machines created by Nature. Unfortunately, our knowledge of biological systems is often limited and we can only try to formulate some assumptions. The problem is that sometimes it is not easy to test an assumption, specially in almost unexplored fields as those inolving neural processes. Recent studies have shown that such test is possible, simulating biological behaviours by means of robots (Chan and Tidwell, 1993; Beer, 1998): we observe a biological system and make some assumptions on its behaviour; then, we build a robot, following the conjectured assumptions, and we see if it behaves as the original system. In this paper we will focus our attention on the cricket phonotaxis (Webb and Scutt, 2000; Webb, 2001): female crickets are able to find their mates recognising and localising the source of the male call song. The hypothesis to be tested is that just a simple neural circuitry, made up of only four neurons, is sufficient to show the complex behaviour of both selecting the correct sound (recognition) and going towards its source (localisation).

2 Neural model

The developed neural system is at the level of membrane potential. Each neuron of the pair of auditory neurons ANL-ANR receives excitatory input from its ipsilateral ear and, in its turn, it excites the ipsilateral motor

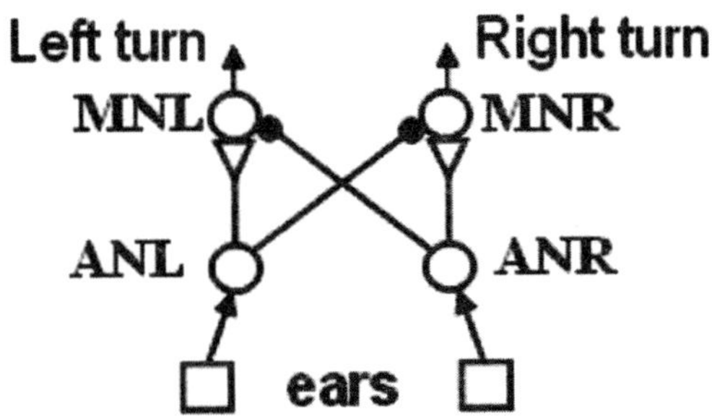

Figure 1. Scheme of the implemented four-neuron model (Webb and Scutt, 2000). From the bottom: the ears, the pair of auditory neurons and the pair of motor neurons. Excitatory synapses are represented by means of open triangles and inhibitory synapses by means of full circles.

neuron and inhibits the contralateral one (see Figure 1). Motor neurons directly drive actuators.
For the sake of simplicity, the model has been simulated using the Simulink MatLab toolbox. Single neurons (see Figure 2) have been represented following the "leaky integrate-and-fire" model (Tuckwell, 1988).

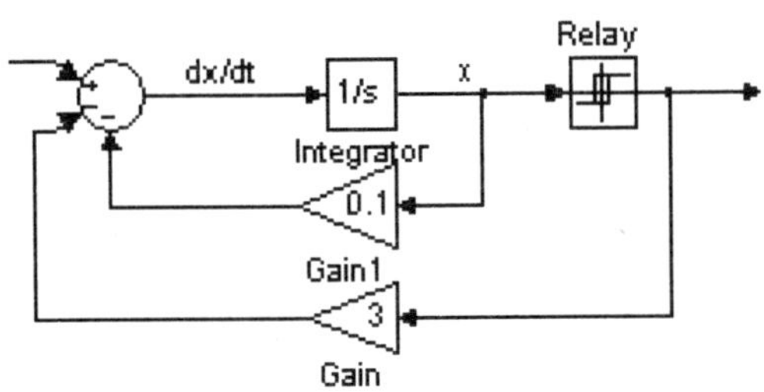

Figure 2. The leaky integrate-and-fire model of a single neuron. The neuron receives input signal from sensorial receptors and/or from the output of other neurons. Then it processes the input by means of an activation function (the relay) and generates the output, which affects the state x in feedback. A term of leakage on the state (-0.1x) is also present.

Figure 3 shows the model of a side of the neural system. Synapse's weight is positive or negative, depending on the fact that there is an excitatory or inhibitory effect on the post-synaptic neuron. Moreover a de-

pression term is added in synapses between AN and MN, i.e. the weight on a synapse decreases while the pre-synaptic neurons are emitting spikes; when the weight reaches a minimum it is recovered to the original value. It follows that only the first few spikes in a burst in the auditory neurons will contribute to generate spikes in the motor neurons.
The implemented neural model is made up by two systems (one for the left side and one for the right side) identical to that of Figure 3. Only two inhibitory synapses must be added (the first one from ANL to MNR and the second one from ANR to MNL) in order to have the complete representation of the scheme of Figure. 1.

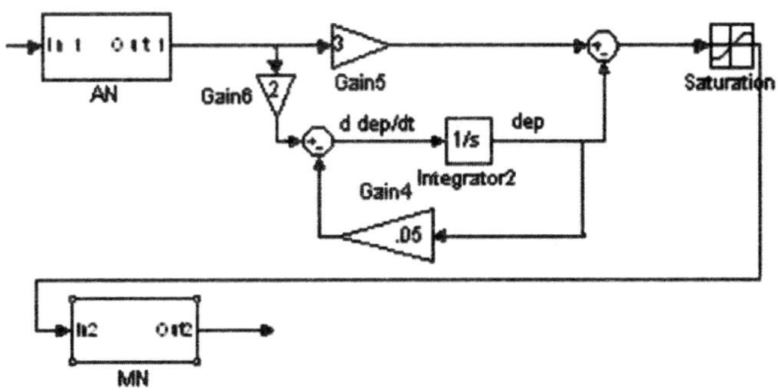

Figure 3. Auditory Neuron (in the upper part of the figure) and ipsilateral Motor Neuron (in the lower part), connected by means of an excitatory synapse, that presents a depression effect.

This simple implemented system allows a rapid localisation of the sound source, dependent on the amplitude of the input signal. If the sound is louder in the left direction, the ANL will be excited more then the ANR. In its turn the ANL will inhibit the MNR and excite the MNL and, consequently, the robot will turn left. Opposite behaviour will be in the case of louder sound in the right direction. However, if the amplitude of sound were the only parameter leading to cricket phonotaxis, no recognition would be possible. For instance, it is worthy wondering what happens if we alter the rate of repetition of the song pattern. Experiments on real crickets (Schildberger, 1984) have shown a striking property of responding more strongly to particular values of this rate. Therefore a careful system parameter tuning is necessary. In particular tuning has been done considering a squared wave as input and taking into account the following points:

- the system has to be able to filter input signals, i.e. it has to show a highly selective behaviour in respect of the intersyllable interval (ISI) for the chirp recognition
- the system has to show that sound recognition does not depend much on the word length (i.e. the duty cycle), according to biological experiments;
- the motor neurons do not have to be excited by the first spike in a burst of the auditory neurons, because it could have been caused by noise.

It is worthy noting that studies on the brain of the cricket have identified a more complex neural structure that affects phonotaxis. In spite of that, the simplified model shown in this paper is a good basis over which more complex implementation may be built.

3 Structure

According to pre-established purposes, we were not interested in a realistic reproduction of the mechanical structure of the cricket: our aim concerned simplicity and speed for realization. Therefore, we chose to use the LEGO MindStorms kit that has become a suitable platform for a broad range of topics. In fact it is characterized by a great flexibility and supports a suite of reusable snap-together sensors (touch, rotation, light, temperature), effectors (motors, lights, infrared (IR) emitters), building blocks, and a programmable control unit (RCX) that can serve as the basis for a wide variety of programming projects (Klassner and Anderson, 2003). The RCX has a 16-MHz CPU (Hitachi H8/3292 microcontroller), 32-kB RAM, and houses an IR transmitter/receiver for sending and receiving data and commands from a desktop PC or from other RCXs. Also programming with Lego is easy: in fact there is a wide set of programming environments and tools in C++, Java and Common Lisp. The robot built in this experiment is endowed with four wheels so it is not bio-inspired certainly (see Figure 4). In order to reproduce phonotaxis, for the robot a microphone receives and amplifies the sound.

4 Algorithm

The sound is a computer-generated song. The standard song consists of syllables holding the length of the chirp constant with intersyllable gaps of the same length as the syllable. It is acquired by the microphone aboard the robot and connected to the audio device of a PC. After the acquisition, signals are filtered with a fourth order low-pass Butterworth filter (Blair and

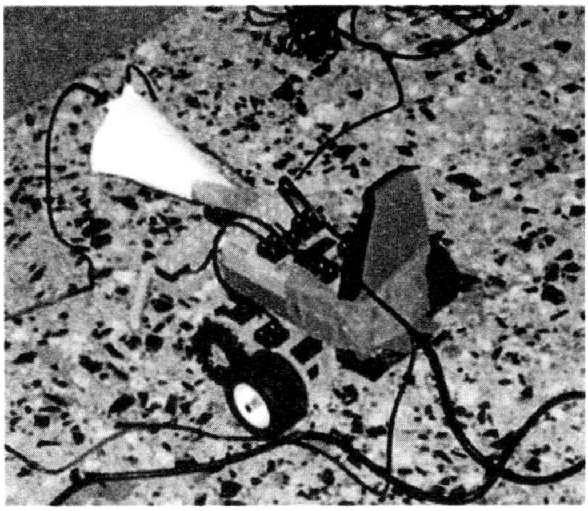

Figure 4. Photo of the Lego robot.

Baraniuk, 2001) (implemented with a MatLab script) and sent as inputs to the auditory neurons. Then we count the number of spikes obtained from the outputs of left and right motor neurons and determine the direction of the sound source, as that corresponding to the side of the motor neuron emitting more. After localization some simple C++ programs drive the Lego motors, sending commands by means of the IR transmitter/receiver connected to a USB port of the elaborating PC. Figure 5 shows the algorithm flow diagram: the initial movement is random, but after the first acquisition and elaboration, if effectively the sound comes from left (i.e. the number of spikes for left motor neuron is higher), the robot will turn left and the ANL will receive the new signal from the new acquisition (Fig. 6 (a)), while ANR receives the old signal; vice-versa if the sound comes from right (see Figure 6 (b)). In this way, even if we use only a microphone, it is possible to simulate a stereo behavior.

5 Conclusion

The main limit of this approach is related to the use of MatLab development environment. That is for the slowness of elaboration especially for the Simulink toolbox, that allows no real time elaboration and, consequently, no real time response by the robot. Another problem concerns the use of only one microphone that leads to displace information from an ear to the other one (or better from a neuron to the other one): this operation can

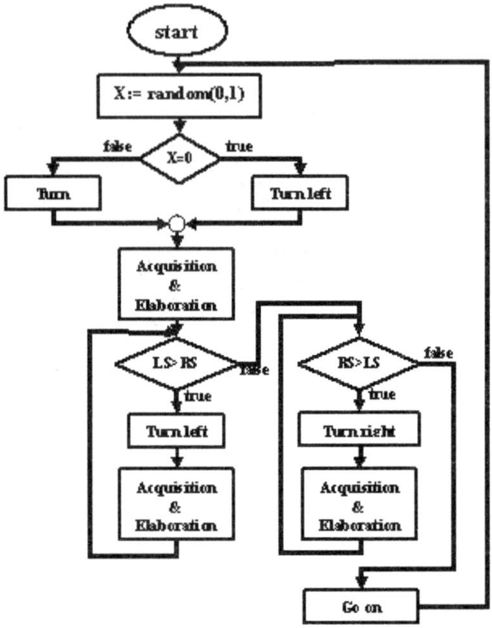

Figure 5. Flow Diagram of the algorithm used to simulate and interpret cricket neural behaviour. LS and RS are respectively the output spikes for the left and right motor neuron.

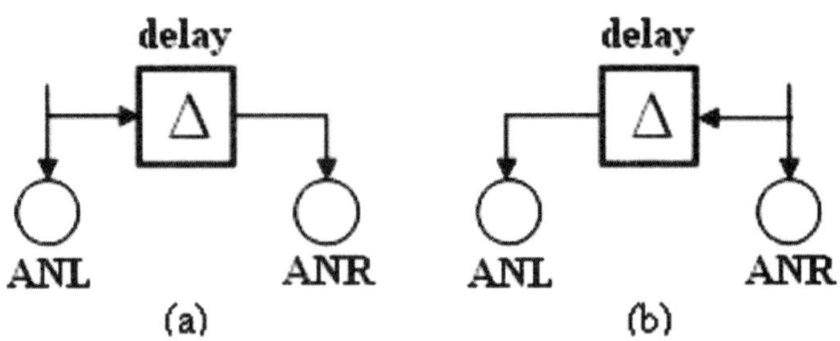

Figure 6. Input signal after a left turn (a) and after a right turn (b)

cause problems related to phase shift and noise-dependent responses. Despite the encountered technical problems, our implementation of the cricket neural system has turned out to be efficient for the tasks of recognizing and localizing the sound source. Moreover the Lego implementation has allowed a quick and cheap robot building.

Bibliography

R.D. Beer. Biorobotic approaches to the study of motor systems. *Current Opinion in Neurobiology*, 8(6):777–782, 1998.

A. Blair and R. Baraniuk. Butterworth filters. *http://cnx.rice.edu/content/m10127/2.1/*, 2001.

K.H. Chan and P.M. Tidwell. The reality of artificial life: can computer simulations become realizations? In *Third International Conference on Artificial Life*, 1993.

F. Klassner and S. D. Anderson. Lego mindstorms: Not just for k-12 anymore. *IEEE Robotics and Automation Magazine*, 10 (2):12–18, 2003.

K. Schildberger. Temporal selectivity of identified auditory interneurons in the cricket brain. *J. Comp Physiol*, 155:171–185, 1984.

H.C. Tuckwell. *Introduction to Theoretical Neurobiology.* University Press, 1988.

B. Webb. Can robots make good models of biological behaviour? *Behavioral and Brain Sciences*, 24(6), 2001.

B. Webb and T. Scutt. A simple latency dependent spiking neuron model of cricket phonotaxis. *Biological Cybernetics*, 82 (3):247–269, 2000.

CNN-based control of a robot inspired to snakeboard locomotion

Peppe Aprile[†], Matthieu Porez[*], Marcus Wrabel[‡]

[†] Dipartimento di Ingegneria Elettrica Elettronica e dei Sistemi, Universitá degli Studi di Catania, Catania, Italy

[†] Université the Bretagne-Sud

[‡] Institut für Strömungslehre und Wärmeübertragung, Technische Universität Wien, Resselgasse 3, 1040 Wien (Austria).

1 Introduction

This paper wants to emphasize the role of analog neural processing structures to realize artificial locomotion in mechatronic devices. The approach presented starts by considering locomotion as a complex spatio-temporal phenomena, modelled referring to particular types of reaction-diffusion nonlinear partial differential equations implemented on a Reaction-Diffusion Cellular Neural Network architecture (RD-CNN). Several examples in literature show the usefulness of this methodology applied to generate and control the locomotion in real-time in a number of different robotic structures as multi-legged or worm-like robots. In this paper we apply this technique, using wave-like solutions, obtained by a ring of RD-CNN cells, to generate and control locomotion in a mechanic wheeled structure that exploits the inertia of two masses.

2 Robot kinematics

The aim of this robot is to use a undulatory movement for robot locomotion. It is based on the paper on snakeboard kinematics (Bullo and Lewis, 2003). The snakeboard consists of two wheel-based platforms upon which the rider has to place each of his feet. These platforms are connected by a rigid coupler with hinges at each platform to allow rotation about the vertical axis as shown in Figure 1.

To propel the snakeboard, the rider first turns both of his feet in. By moving his torso through an angle, the snakeboard moves through an arc

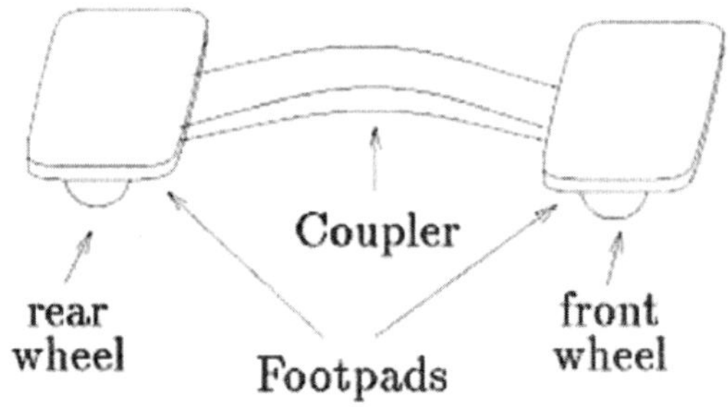

Figure 1. Schematic kinematics of a snakeboard.

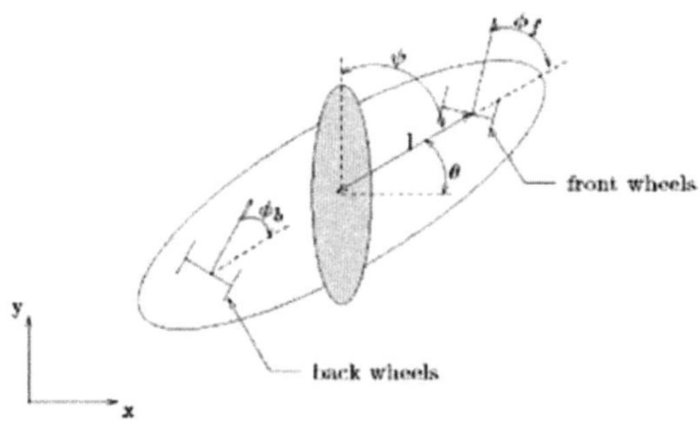

Figure 2. Kinematics of the robot.

defined by the wheel angle. The rider then turns both feet so that they point out, and moves this torso in the opposite direction.

By continuing this process the snakeboard may be propelled in the forward direction without the rider having to touch the ground. Figure 3 represents the kinematics of robot that has been realized during the workshop.

The rider is replaced by a mass positioned on the mass center of the snackboard. The inertia of the mass must be sufficient to move the robot. The back wheels, the front wheels and the mass are controlled by the angle between the set of wheels and the average fibre of the rigid body. The position of the robot is defined by the position of mass center.

Figure 3. Photo of the robot.

3 Implementation of the control strategy

The system is constituted by two steering axes with non-actuated wheels and a propulsion mechanism based on a rotating beam with two masses on each extremity. The idea is to realize the control of the three actuators using the characteristic slow-fast dynamics of autowave fronts propagating on a ring of n-cells (while a single cell control a specific actuator) and co-ordinate the whole movement by adding or subtracting cells in the ring. The whole locomotion control can be divided in three steps:

1. The two steering axes move almost synchronous, one rotates counterclockwise and the other one rotates clockwise.
2. The beam from its initial position, that we can consider almost perpendicular to the locomotion direction , rotates counterclockwise, slowly accelerating and quickly decelerating.
3. Steps 1 and 2 are repeated in the opposite direction (see Figure 4).

Let us consider the CNN approach control strategy: first we can use a basic ring of two cells to generate two different phases required to control independently the axes and the beam, assuming that the steering are actuated by the same cell in opposite phases. Furthermore we can assume that the values 1 and -1 of the cell output represent the steady state, whether for each axis, or for the beam. Then in the same way the value 0 represent the perpendicular position respect the whole body orientation. In particular, in the case of beam control we can use the characteristic shape of the output slow-fast dynamics to realize a slow acceleration of the masses and the consequently quickly deceleration to set the specific requested motion.

The same wavefront with a specific delay due to the connection parameters or to the number of cells in the ring between the two cells used for

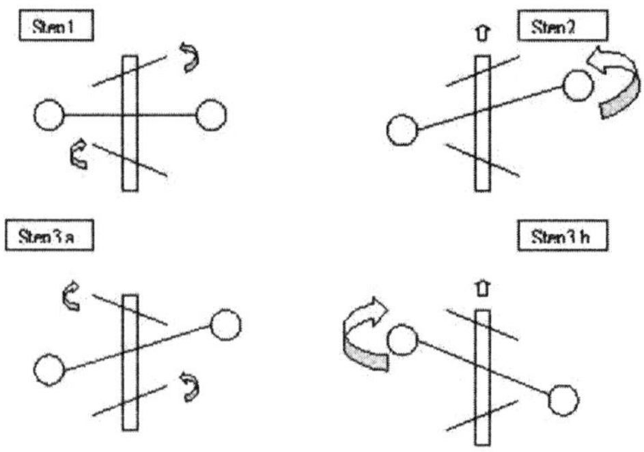

Figure 4. Sequence of steps in the control algorithm.

the control propagates through the ring and actuate the two steering axes as shown in Figure 5.

In this way (by opportunely synchronizing the wavefronts by increasing the number of cells of the structure), we can obtain the desired control of the actuators.

For instance a possible solution is shown in Figure 6. In this implementation we considered the two axes controlled each one by a different output variable of a single cell, for example Y1 and Y2. This is due to the fact that the phase delay between the two variables is very low. In Figure 6 the outputs of the two cells considered are shown: the dotted lines represent the outputs Y1 (red) and Y2 (blue) of the first cell, while the continuous lines show the second cell behavior in the time domain. For an example the output Y1 of the second cell, the variable Y2 of the second cell and the output Y1 of the first cell could control the second steering axis, the second axis and the beam movement, respectively.

Bibliography

F. Bullo and A.D. Lewis. Kinematic controllability and motion planning for the snakeboard. *IEEE Transactions on Robotics and Automation*, 19(3): 494–498, 2003.

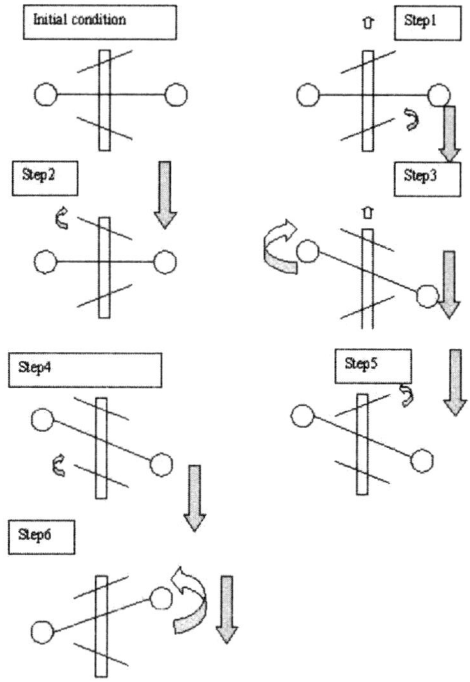

Figure 5. Control sequence for actuating the robot.

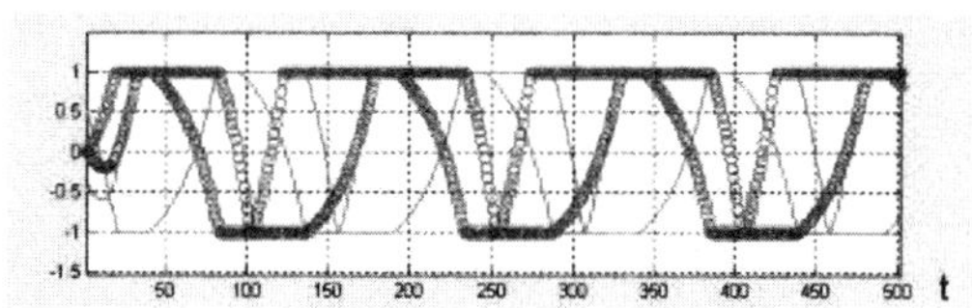

Figure 6. Trends of the variables controlling the three degrees of freedom of the robot.

Cooperative behavior of robots controlled by CNN autowaves

Paolo Crucitti, Giuseppe Dimartino,
Marco Pavone and Calogero D. Presti

Scuola Superiore di Catania, Catania, Italy

Abstract In this paper we will discuss in some details how the wave-based approach can be applied to roving robots. Two slightly different implementations are shown and experimental results obtained are discussed. The wave-controlled robots have only local information coming from the on-board cameras, but a simple cooperation strategy shown in the following allows to overcome the limits of local information.

1 Introduction

In autonomous robot design the most severe problems lie on how to deal efficiently with vision and path-planning decisions. In fact, both image elaboration and path-planning decision are time consuming processes that may require long elaboration time. On the other hand, as shown in previous papers, the availability of an integrated circuit provided with both a CNN circuit (Chua and Yang, 1988b,a; Chua and Roska, 1993) and a CCD matrix allows to perform both image acquisition and elaboration in parallel, so in a very fast way. Therefore, our aim is to exploit the CNN capabilities to take path-planning decisions. This is possible thanks to a particular class of phenomena that can occur on a CNN medium: propagation of autowaves (Arena et al., 1999).

In this paper we will discuss in some details how the wave-based approach can be applied to roving robots. Each robot is equipped with an on-board camera. Moreover the robots cooperate to reach the target. Since image processing techniques based on CNN are well known, we have rather focused our attention on the path-planning algorithm. Therefore, image acquisition and obstacle and target recognition are realized with classic digital techniques. In more details, image elaboration is aimed to distinguish between obstacles and targets resting on their different dimensions; they are simply black Lego™ pieces placed on a totally white floor.

2 Path planning through autowaves

The main idea is to consider the environment (or, rather, its image as acquired by CCD) as an active media in which autowaves can propagate (Adamatzky et al., 2004). In particular, obstacles are assumed as attractive autowaves generators, while target as repulsive autowaves generators. Obviously, autowaves travel on two different CNN, the former for obstacles, the latter for targets; they propagate until *robot sensitive pixels*, placed around the robot, are reached. Robot sensitive pixels are a small number of pixels that allows to estimate obstacles and targets direction with respect to the robot. Moreover, since autowaves speed is constant, distance is simply estimated. Finally, non-interference property guarantees that just the closer obstacle (or target) is detected, since its autowave prevails.

Given direction and distance information, it is possible to take suitable actions in order to avoid obstacles and reach the target.

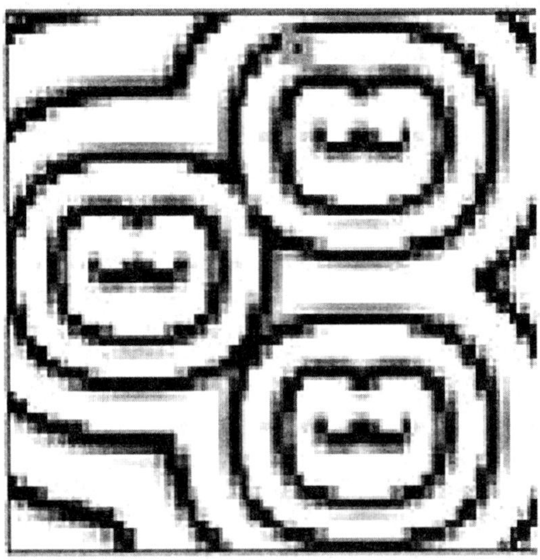

Figure 1. Autowaves generated by three obstacles.

3 Robots structure

To validate the approach stated above, we have implemented two different robots in Lego MindStorms^{TM}. This simple and flexible system is well-suited for a fast implementation, avoiding many mechanical problems arising in the design of a robot.

3.1 Common features

Both robots are provided with a tower on which a webcam is placed and with a target signal rod (whose purpose is described later) in the back part. As regards the mechanical structure, each robot has three driving wheels per side. This solution turned out to be optimal in terms of rotation precision and occupied space. Furthermore, each robot is able to change its motion direction thanks to a differential drive. All the computations (image processing and autowave simulation) are performed on a PC; the data exchange between the microprocessor and the PC is realized via infrared communication. Another important common feature is that each robot is controlled only on the basis of local information arising from the on-board camera.

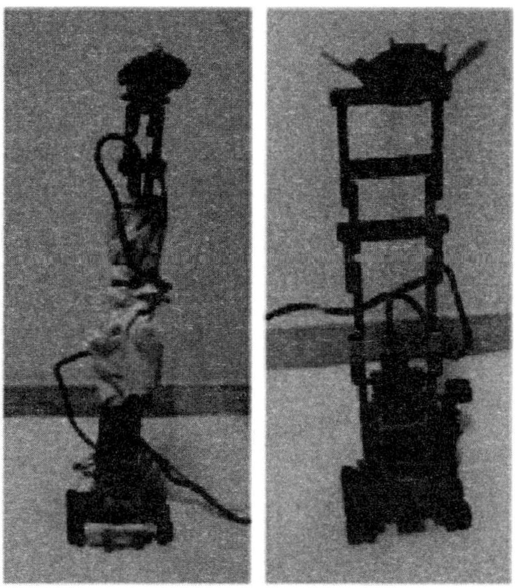

Figure 2. Mercurius (on the left side) and Shybot (on the right side)

3.2 Mercurius robot

The first robot is called Mercurius and is shown in Figure 2. The dimensions of Mercurius are:

- Width: cm 17.5
- Length: cm 15.5
- Height: cm 74

Rotation angle is measured through a differential gear connected to a rotation sensor. This information allows both to know the swerve from the *preferential direction* and to perform feedback rotations.

Sensitive pixels are shown in Figure 3. Their geometrical disposition is aimed to approximate a semicircumference in order to detect as better as possible an incoming autowave (e.g. an incoming autowave from north-west will activate pixel 2). We noted that none of the pixels revealed to be dominant.

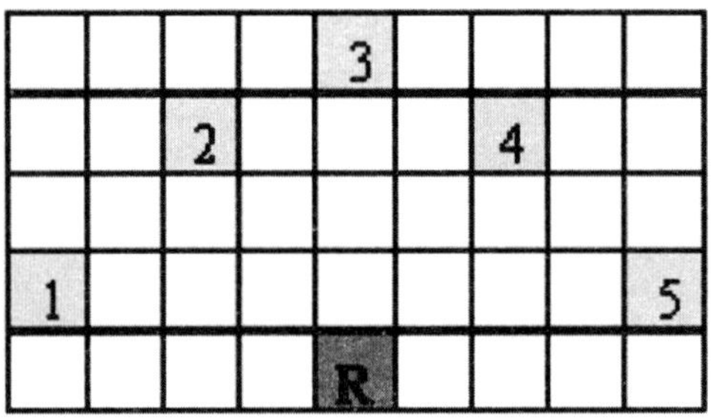

Figure 3. Mercurius sensitive pixels

At the beginning, the robot is aligned with a given initial preferential direction; Mercurius tends to maintain this direction until a target is detected, then the new preferential direction becomes the estimated target direction.
Let us define:

- α: the angle between robot's north and the preferential direction
- γ: angle between robot's north and the estimated obstacle direction
- β: angle between robot's north and the normal to the estimated obstacle direction

Obviously, $\beta + \gamma = \frac{\pi}{2}$, so we defined γ angle only for convenience.

α and β angles are such that:

$$-\pi \leq \alpha \leq \pi$$

$$-\frac{\pi}{2} \leq \beta \leq \frac{\pi}{2}$$

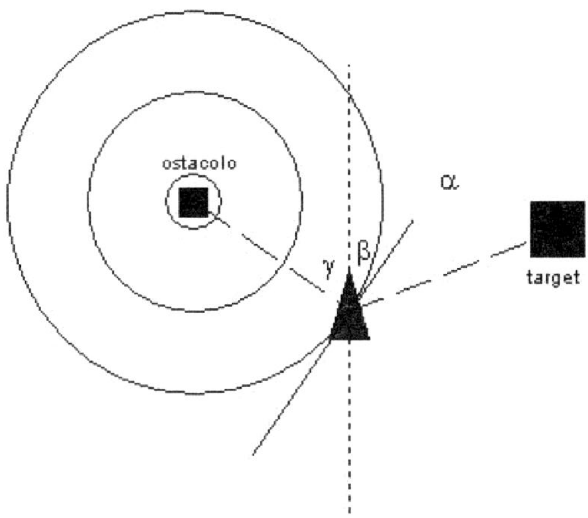

Figure 4. Definition of angles α, β and γ.

The angle γ is calculated basing on pixels activation because of incoming autowaves. If we assign at each pixel a vector in a coordinate frame in which the robot is placed in the origin, the angle γ is obtained by the sum of all vectors corresponding to activated pixels. On the other hand, α angle is calculated resting on rotation sensor. As far as target is concerned, the angles α, β and γ are obtained in the same way.
As stated above, when a target is detected, its direction becomes the new preferential direction; therefore, the rotation sensor is re-initialized.
Then, the fundamental idea, in the case of a very close obstacle, is to align the robot along the direction normal to the estimated obstacle direction (according to information given by β), in the semi-plane, of course, in which the preferential direction lies (as α angle dictates). In case in which the obstacle is far enough to ensure safe walking, the robot will go forward and then turn toward the preferential direction. Thus, the robot will tend to mantain a safety distance away from each obstacle.

This algorithm is based on the *a priori* knowledge (at least approximated) of target direction: if it is unknown, there are not guaranties that the robot can reach the target.

3.3 Robot Shybot

The second robot is called Shybot and is shown in Figure 2. This robot is characterized by the following size:

- Width: cm 16;
- Length: cm 13;
- Height: cm 55.

As regards the CNN of Shybot it can be represented as in Figure 5. It possesses seven sensitive pixels (numbered from 1 to 7) that are symmetrically placed among right and left sides. If obstacle autowaves first intersect a left side pixel (1, 2 or 5), the robot will avoid collision with the obstacle by turning right. A specular behavior is provided for the case in which the right side pixels (3, 4 or 6) are first intersected. No rotation will occur if the first hit pixel is the back one (7).

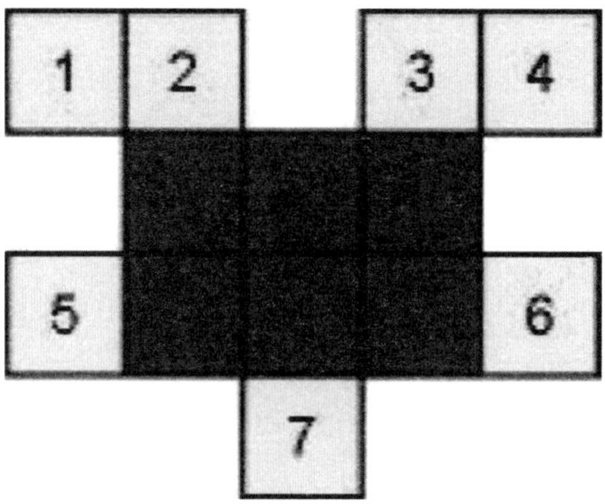

Figure 5. Sensitive pixels of Shybot.

Also the rotation angle is obtained as a function of the stricken sensitive pixel: wider rotations in the case obstacles are in front of the robot and lighter rotations in the case obstacles are in a lateral position. After the determination of this rotation angle, a random noise is added in order to avoid problems related to local minima (e.g. high symmetry).

As far as the advance manoeuvres are concerned, the robot behavior is a function of the obstacle distance. If obstacles are near, Shybot will draw back (before rotation), while if obstacles are far away, Shybot will go forward (after rotation) for a tract of length proportional to obstacles distance.

If more pixels are intersected at the same time, rotation angle and advance will be determined averaging on the values related to the stricken pixels.

Obviously, the reaction to targets is the opposite to that of obstacles. The robot will turn in the direction of the nearest target and go towards it. If both obstacles and targets are present on the field of vision, the robot will behave in two different ways corresponding to two different cases:

1. Obstacle autowaves arrive first: it means that there is at least an obstacle between the robot and the nearest target; the robot will wait for the target autowaves to arrive, it will go round the obstacle and then it will move with the purpose of keeping the target inside its field of vision.
2. Target autowaves arrive first: it means that there is no obstacle between the robot and the nearest target. The robot will not wait for the obstacle autowaves and it will go towards the target.

Some particular cases have been taken into account for the movement strategy:

- if obstacles are too near respect to the size of the robot, a wall of obstacle will be built in order to avoid that the robot trying to pass between them;
- if there are no obstacles and no targets in the field of vision, the robot will advance and rotate randomly.

4 Cooperative behaviour

Many biological systems show that cooperation between individuals of the same species plays a fundamental role in pursuing vital activities such as food search (Bonabeau et al., 2000).
In our work we implemented cooperation in a simple way: when a robot sees the target it points out its sighting to the rest of the herd by means of a target signal rod, whose final part shows the same shape of the target (Figure 6). In this work we built only two cooperating robots. Nevertheless the cooperation can be easily extended to an indefinite number of robots. More the cooperating robots will be used, better will be the cooperative behavior. The target rod is lowered when the target enters the field of vision and raised when the target exits. Consequently, if a robot A is far away from the target, but it sees another robot B whose field of vision contains a target (real or fictitious), the robot A will follow the robot B until a real target is reached.

Power of cooperative behaviour is increased by the fact that robots are similar but not identical. It means that they share the same objective and they attain their goal in a similar way, but their differences cause a different fitness to the environment. The herd adjusts to an environment in the sense

Figure 6. The target signal rod of Shybot.

that the fittest robot finds the best route to reach the target and guides the others towards it. This property lends stability to environmental changes.

5 Discussion

Experimental results validated the autowave based approach for path-planning decisions and, moreover, the cooperation strategy. Autowave based approach, as stated above, allows real-time image processing and decision planning and guarantees robustness in respect of noise presence; on the other side cooperation reduces considerably the time spent for target searching. Then, since autowave based path-planning in conjunction with CNN image processing show clear advantages compared to traditional path-planning algorithms, next step will be the integration of all the functionalities on CNN chips. Furthermore, as far as cooperation is concerned, next step will be the employment of a greater number of robots, in order to obtain more interesting behaviour in terms of complexity and self-organization.

Acknowledgement. Part of this project had been developed in Catania, as closing work for the course of Bio-robotics held by Professor P. Arena at the Scuola Superiore di Catania. We thank Giuseppe Aprile, Giuseppe Mattiolo and Giovanni Puglisi for their useful collaboration in the project.

Bibliography

A. Adamatzky, P. Arena, A. Basile, R. Carmona-Galan, B. De Lacy Costello, L. Fortuna, M. Frasca, and A. Rodriguez-Vazquez. Reaction-diffusion navigation robot control: from chemical to vlsi analogic processors. *IEEE Trans. on Circuits and Systems I*, 51 (5): 926–939, May 2004.

P. Arena, L. Fortuna, and M. Branciforte. Algorithms to generate and control artificial locomotion. *IEEE Transaction on Circuits and Systems I*, 46 (2):253–260, 1999.

E. Bonabeau, M. Dorigo, and G. Theraulaz. Inspiration for optimization from social insect behaviour. *Nature*, 406:39–42, 2000.

L. O. Chua and T. Roska. The CNN paradigm. *IEEE Transactions on Circuits and Systems*, 40:147–156, 1993.

L. O. Chua and L. Yang. Cellular Neural Networks: Applications. *IEEE Trans. on Circuits and Systems I*, 35:1273–1290, October 1988a.

L. O. Chua and L. Yang. Cellular Neural Networks: Theory. *IEEE Trans. on Circuits and Systems I*, 35:1257–1272, October 1988b.